I0043546

A.O. Kittredge

The metal worker, essays on house heating by steam, hot water and hot

air

With introduction and tabular comparisons

A.O. Kittredge

The metal worker, essays on house heating by steam, hot water and hot air
With introduction and tabular comparisons

ISBN/EAN: 9783743467095

Manufactured in Europe, USA, Canada, Australia, Japa

Cover: Foto ©berggeist007 / pixelio.de

Manufactured and distributed by brebook publishing software (www.brebook.com)

A.O. Kittredge

The metal worker, essays on house heating by steam, hot water and hot air

THE METAL WORKER

Essays on House Heating

BY STEAM, HOT WATER AND HOT AIR

WITH INTRODUCTION AND
TABULAR COMPARISONS.

ARRANGED FOR PUBLICATION BY

A. O. KITTREDGE,

Editor of *The Metal Worker, Carpentry and Building*, etc.

Author of the "Metal Worker Pattern Book," "Practical Estimator," etc. .

NEW YORK:
DAVID WILLIAMS.
1891.

Copyright, 1891,
By DAVID WILLIAMS.

CONTENTS.

PRIZE WINNERS.

———:·:———

COMBINATION SYSTEMS:

    Steam and Warm Air—JOHN MILLAR,

             Toronto, Ont.   Special Prize.

    Hot Water and Hot Air—A. R. BRINK,

             Red Wing, Minn.   Special Prize.

STEAM HEATING SYSTEMS:

    First Prize—E. P. WAGGONER,

             Syracuse, N. Y.

    Second Prize—HARELL STEAM HEATING CO.,

             Dunkirk, N. Y.

HOT-WATER CIRCULATING SYSTEMS:

    First Prize—RICHARD SWALWELL,

             Vancouver, B. C.

    Second Prize—JOHN HOPSON, JR.,

             New London, Conn.

HOT-AIR SYSTEMS:

    First Prize—ANSON W. BURCHARD,

             Danbury, Conn.

    Second Prize—JAMES A. HARDING,

             New York.

INTRODUCTION.

In the fall of 1888 *The Metal Worker* announced a series of Prize Competitions in House Heating, which subsequently became famous the country over. The floor plans and elevations of a dwelling house, which had been the subject of previous competitions in the periodical known as *Carpentry and Building*, were used as the foundation. The elevations presented were one-eighth inch scale, and plans of the same scale were printed on separate sheets adapted to receive at the hands of the competitors lines indicating the position of heater, runs of pipes, location of radiators, registers, etc., thus insuring uniformity in presentation. Three systems of heating were recognized in the competitions—viz: steam circulation, hot water circulation and hot air. Provision was also made for the consideration of combination plans. The results of the contest, as viewed from the number and excellence of the essays submitted, were a gratifying success, while the publication in *The Metal Worker* of the prize essays, together with selections from the best efforts that received no prizes, covering as it did a period of several months, both pleased and instructed its large constituency of readers.

The essays presented in the following pages are reprinted from the columns of the paper referred to, and are offered in the present form, in the belief that they will prove useful to all

GENERAL VIEW OF HOUSE FURNISHING THE BASIS OF THE METAL WORKER ESSAYS ON HEATING.

who are concerned in the heating of houses by any of the systems above named. In addition, there are included summaries derived from very careful study of the competitive efforts. These summaries, both in the matter of cuts and tables, serve to show in brief the results of many minds working upon the same problem, and cannot fail to be of particular interest and value to all in the trade.

FRONT ELEVATION, EAST.—SCALE, 1-16 INCH TO THE FOOT.

According to the conditions which were laid down, the contestants were to show the arrangement of apparatus and indicate the runs of pipes, to present full descriptions of the work, with the amounts of heating surface they would use in each of the several rooms, together with the reasons for each step taken. A detailed estimate of cost was also required. Concerning the latter it should be remarked that in arranging the details of these competitions it was thought that better

results would be obtained if the limit of cost was left largely to
the discretion of the individual competitors.   In most instances
in active business the heating engineer is required to comply
with the preconceived notions of the house owner, so far as cost
is concerned.   He is very often required to show how much he
can do for a given sum.   Accordingly, some parts of the work
are necessarily slighted.   In this case the competitors were

SOUTH SIDE ELEVATION.—SCALE, 1-16 INCH TO THE FOOT.

requested to look the house over, prepare their own specifications as they would like to see the work done, and to submit
the bill of particulars. The judgment of the individual contestants in this matter—quite naturally—varied, and therefore
there was found in the efforts submitted a very wide range of
costs. Coming as they do from various parts of the country
these essays, in a sense, present a bird's-eye view of the practice
of heating engineers in general. They constitute a most useful text book for the instruction of any who are beginning to

give attention to the general question of house heating, while to men of some experience in the business they will be found to contain many valuable suggestions—for example, in the matter of proportioning surfaces, in minor details of construction and in the practice of estimating.

The essays submitted are arranged under four heads. 1, Combination Systems; 2, Steam Systems; 3, Hot-Water

REAR ELEVATION, WEST.—SCALE 1-16 INCH TO THE FOOT.

Systems, and 4, Hot-Air Systems. In the first, two essays are presented, one employing a combination of steam and hot water and the other a combination of hot water and hot-air. In the steam heating section four essays are contained, showing quite a range of practice. In the hot-water section three essays are given, also indicating a wide range of practice, while in the hot-air section six essays are presented, likewise indicating that there are many different

ideas of the proper heating of the house in question by the use
of hot-air furnaces.   In the steam and hot-water sections it
will be noticed that some of the writers have adopted the direct
system of heating ; others have preferred the indirect system,
and still others have used a combination of direct and indirect.
It is believed, taken altogether, that the fifteen essays here
submitted present a better idea of current practice in house
heating than can be found anywhere else.   The conclusions
and comparisons presented in the several supplementary
chapters add still further to the general fund of information.

A. O. K.

New York, April, 1891.

I. COMBINATION SYSTEMS.

STEAM AND WARM AIR.*

BY JOHN MILLAR.

This system combines good ventilation with any desirable temperature during the coldest weather ; is easily controlled, and economical in fuel.

The pure air is introduced through the outside air duct (as shown on basement floor plan, Fig. 1), conducted into a pit under the furnace. From there it ascends, and is heated in its passage between dome and radiator of furnace, passing upward through the warm-air pipes to the different apartments to be heated. This pure air rises to the ceiling, and the heavier or impure air, falling to the floor, is carried off through the fireplace flues, and, in the rooms that are not provided with fireplaces, through the registers that are placed in the walls and connected with flues to be built for ventilation, as described further on and shown on the plans. *Introduction of Pure Air.*

This system of drawing fresh air from the outside and passing it into the building heated is done without any extra firing or danger of frozen pipes, as often occurs where the indirect system of steam or hot water is used.

The warm air supply from this furnace will be sufficient to heat the house in ordinary winter weather. When the thermometer drops down in the neighborhood of zero, the reserve or steam power can then be utilized by firing up the furnace a little heavier, thus giving a good supply of heat all over the building during the most severe weather. By having a warm-air register and steam radiator in each room the heat can be turned off one or both, thus lowering or raising the temperature to suit the occupant. *Warm Air Supply.* *Auxiliary Steam Heat.*

The draft of the furnace is controlled and regulated by a steam diaphragm. This can be set to carry one, two or three pounds of steam, or more if desired, at the will of the person attending it. *Control of Draft.*

It is not only desirable that a heating apparatus should be a good heater, but it should also be economical. This furnace has a wrought-steel dome and radiator. The direction of the draft is to the top of the dome, then into the radiator that surrounds the dome, passing around this before entering the *Economical Features.*

* From *The Metal Worker*, March 2, 1889.

smoke-pipe, thus retaining the products of combustion for
some time before passing up the chimney.

A small fire with front draft closed and check draft in
smoke-pipe open will throw off sufficient heat to warm a
house in moderate weather, the light steel plate radiating the
heat very quickly. Heavy firing is not required, even in the
coldest weather.

Sheet-Steel
Furnace.

It is a well known fact that sheet steel will radiate heat
quicker and with a smaller consumption of fuel than cast
iron, and as the fuel does not come in contact with the steel
dome and radiator its durability is assured. The fire-pot being extremely heavy, it does not get red hot, this being objectionable in furnace heating; so that with a good large fire in
furnace and front dampers tightly closed the fire is held for a
good many hours with a moderate consumption of fuel.

VENTILATION.

Ducts for
Pure Air

To ventilate a building properly some means must be
adopted to furnish a good supply of pure warm air. This I
propose to do by building a pure-air duct from the north side
of the building (see Fig. 1), 14 x 30 inches, inside measurement, to connect with a duct built under the cement floor in
the cellar; this duct to be built of brick, plastered smooth
on the inside; the top to be covered with stone slabs, and to
connect with a pit under the furnace; this pit to be same
diameter as lower ring of furnace and 14 inches deep, built
of brick and plastered smooth on the inside. Another duct
to be built and connected with cold-air register in the hall
near the front door, 10 x 20 inside, constructed in same way
as the outside duct, and connected with pit under the furnace.
These ducts to have each a damper, with chains so attached
to them that they can be operated from the first floor. Either
of these ducts can be used as desired.

Exhaust of
Impure Air.

The pure air being introduced, the impure air must now
be got rid of. The fireplace flues in the principal rooms will
answer the purpose well; the rooms that are not so provided
to have brick ventilating flues, constructed as shown on plan.
Kitchen, one flue, 8 x 8, inside measure, plastered smooth,
built alongside smoke flue, and to have one 8 x 12 ventilating
register placed in it near the wall for the purpose of carrying
off the steam and smoke from the kitchen.

No. 4. Bedroom over kitchen, Fig. 3, to have one 8 x 10
register, placed in same flue near the floor.

Smoking-room in attic, Fig. 4, to be provided with two 8 x 10
registers, one near the ceiling, to carry off fumes of tobacco,
&c., and the other near the floor, to be connected with a brick
flue 8 x 8 inch, as shown on plan.

Cellar Plan.—Scale, 1-12 Inch to the Foot.

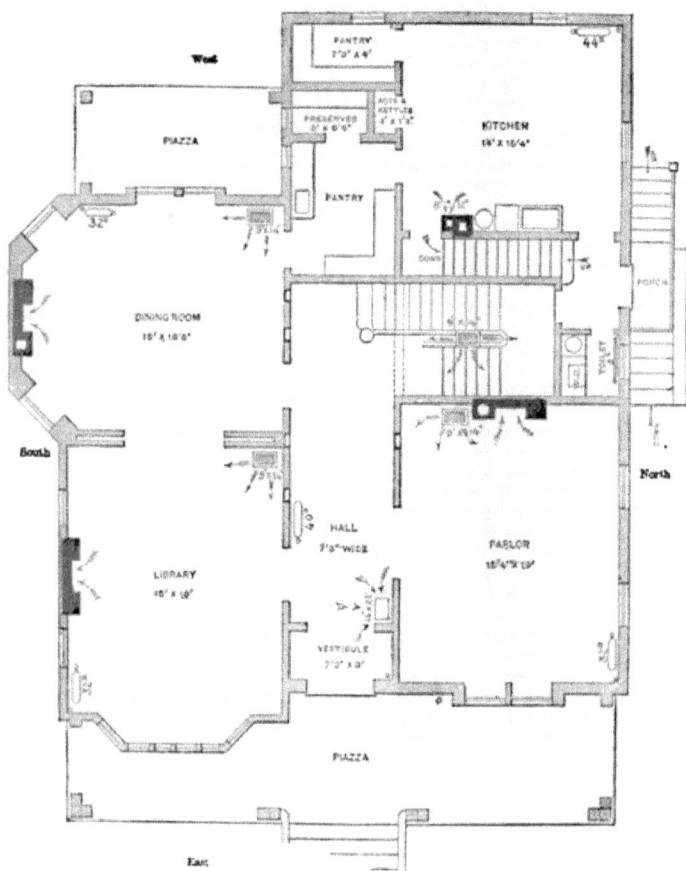

First-Floor Plan.—Scale, 1-12 Inch to the Foot.

Playroom, Fig. 4, to have one 8 x 10 register near the floor, connected with flue.

Chamber, Fig. 4, to have one 9-inch register, connected with a 7-inch galvanized pipe, running across to a flue, to be provided as per plan.

HEATING.

The warm-air registers on the first floor, Fig. 2, are placed in the floors. This is best place for them. The warm air can flow directly through the pipes to registers without any friction, there being no contraction of the pipes such as would be the case were registers put in the base of walls.

The warm-air pipes to be taken from the side of the furnace at the top, and all collars to be placed on a level (top side), irrespective of size. **Pipes.**

The sizes of the warm-air pipes and registers are determined upon according to exposure of rooms, distance from furnace, whether on first, second or third floors, giving the advantage, if possible, to north and west rooms on first floor. The register in dining-room, Fig. 2, is placed at some distance from the furnace, to allow the placing of a sideboard between the doors from hall and pantry, this being the only part of the room in which this article of furniture can be placed; but, as this is a south room and has the steam radiator to assist in heating it, there will be no difficulty in heating this room. **Pipes and Registers.**

First-floor pipes are 9-inch; second floor, 8-inch, with the exception of the bathroom, which is 7-inch—it being a short pipe, the bathroom will very easily be warmed. Pipe to chamber on attic floor is 7-inch, because it is a short one, and has about 25 feet of vertical pipe that will draw well. The pipe to smoking and play rooms is made 8-inch, because it is a longer pipe and has larger rooms to heat. There is no rule to guide a man in placing a furnace and pipes better than experience, and this work has been laid out based upon that. **Size of Warm-Air Pipes.**

Steam radiators are placed in the coldest corners, the cold air from the windows in these corners being warmed in its contact with radiators. These radiators will make the warm-air registers work better, for it is much easier to draw warm air through the registers into a partially heated room than a cold one. The 14 x 22 cold-air register in hall, Fig. 2, close to the front door will draw off the cold air from the floor, making it very much easier to heat the hall. A good circulation can be secured by the use of this register, the air supply from this register being as pure as that from the outside. In very severe weather the outside duct can be closed say two-thirds, and with the other third and air from the hall register there will be sufficient to supply the furnace. **Radiators.** **Cold-Air Register.**

WARM-AIR SPECIFICATION.

Casing. Furnace to be portable and double-cased. Outside casings to be made of galvanized iron, 24 gauge ; inside casings to be good charcoal tin plate, 28 gauge, with a space of 1 inch between **Base.** outer and inner casings. The base and ash-pit to be of heavy **Fire-Pot.** cast iron. The fire-pot to be cast iron, 22 inches in diameter, 17 inches deep, and to weigh not less than 400 pounds. Diam-**Dome.** eter of grate, 20 inches. Dome to be made of wrought-steel plate, best quality, No. 12 gauge ; top end steel plate, No. 10 gauge, and strongly riveted in. Dome to be surrounded with a wrought-steel radiator, No 16 gauge. Furnace to be supplied with one cast-iron vapor-pan, dust-flue, poker, shaker and scraper. Joints to be made perfectly secure from dust and gas escape.

Radiator Finish. Each radiator to be supplied with 1 nickel-plated radiator valve with union and round wood handle; also, 1 automatic air valve, nickel-plated. Coil in toilet-room to have 2 valves and 1 automatic air valve.

Flow-Pipes. The main steam-pipes from boiler to be laid with a fall of 1 inch in 10 feet. Branch pipes to be taken off the top of mains. Great care to be taken that there be no traps in the pipes.

Size of Steam-Pipes. Main steam-pipes to be 2 inches in diameter. Risers and branches to all radiators to be $1\frac{1}{4}$ inches.

Branch pipe to coil in toilet-room to be 1 inch in diameter. These pipes to be laid to all radiators on the one-pipe system.

Return-pipe from wall coil to be ¾-inch pipe.

Where steam-pipes run through brick walls a thimble of galvanized iron $\frac{1}{4}$ inch larger than the pipe to be used, to allow for a free expansion of steam-pipes.

Floor and ceiling plates to be provided for all pipes in rooms and halls. These to be nickel-plated.

Return-Pipes. The return-pipes to be placed under the cellar floor, with a proper fall toward the boiler. These pipes to be 1¼ inches in diameter, and well covered with sifted coal ashes.

The drop-pipes connecting main steam with main return to be ¾ inch in diameter.

All piping to be of the very best make. Fittings to be cast iron.

Water-Supply. A proper connection will be made with the water system of the house, so that the boiler can be supplied with water conveniently.

Radiators and all exposed pipes in rooms and halls to be bronzed.

Pipe Covering. All pipes exposed in basement to be covered with asbestos, hair felt ¾ inch thick, and covered with canvas neatly and strongly sewn on.

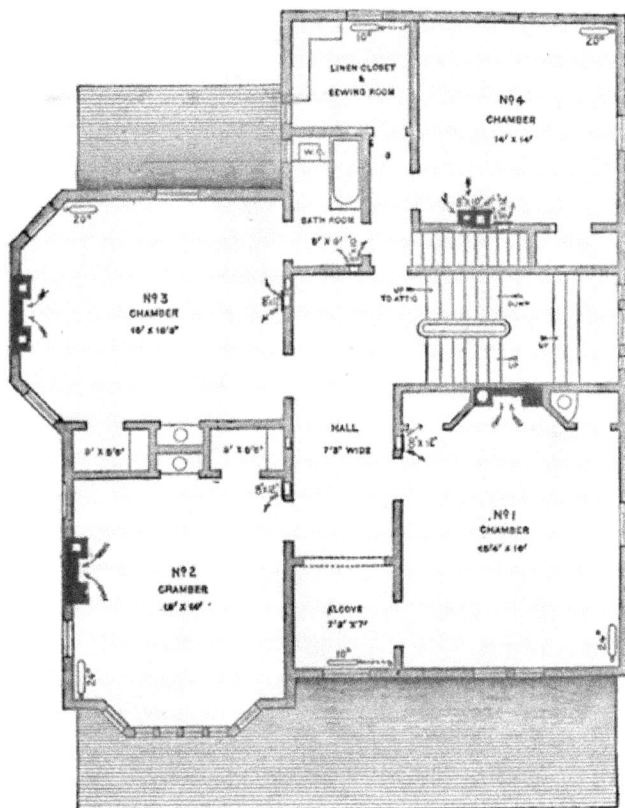

LINEN CLOSET
&
SEWING ROOM

N⁰4
CHAMBER
14' x 14'

BATH ROOM
8' X 8'

N⁰3
CHAMBER
16' X 16'0"

UP
TO ATTIC

DOWN

HALL
7'8" WIDE

N⁰1
CHAMBER
16'6" X 16'

N⁰2
CHAMBER
16' X 16'

ALCOVE
7'8" X 7'

Second-Floor Plan.

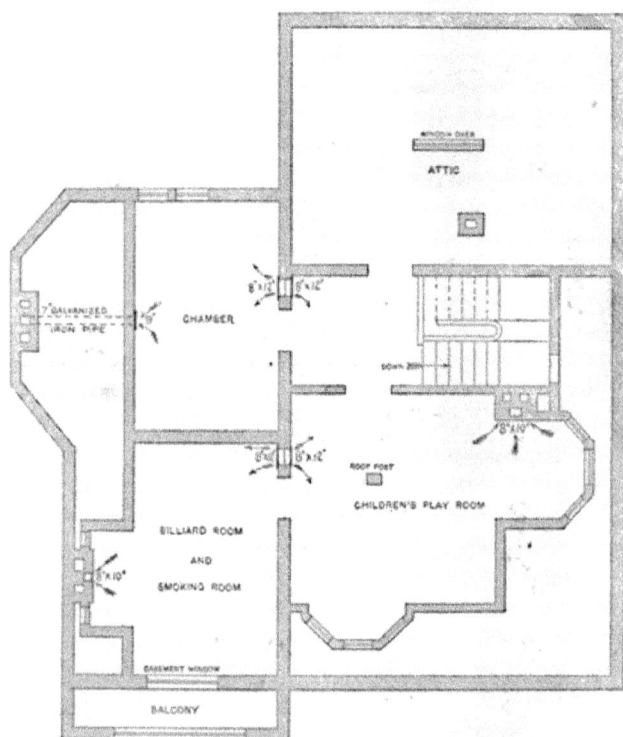

Attic Plan.—Scale, 1-12 Inch to the Foot.

The blow-off pipe to be 1¼ inches in diameter and to be properly connected with drain. *Blow-off Pipe.*

The materials used and workmanship employed to be the best of their respective kinds, and the apparatus left in a clean and tidy shape, all ready to fire up. *Materials.*

The purchaser to build smoke and ventilating flues, pit under the furnace, air ducts, and do all cutting for registers and pipes (the sheet-iron laths to be furnished by the contractor for the plastering), and to furnish good coal and careful attendance.

I will agree to fulfill contract as herein specified for the sum of seven hundred and nine dollars and eighty cents ($709.80). *Cost.*

Freight charges, traveling and other expenses incurred, if this work is to be done outside of the city, not included in above price. *Freight.*

GUARANTEE.

I will agree to heat the building with the apparatus as herein specified, without forcing, in zero weather, up to 70° in the rooms on first floor and children's playroom, halls and chambers to 65°, bathroom to 75°.

The warm-air pipes to be made of the best IX tin plate. *Warm-Air Pipes.*
Smoke-pipe to be 9 inches in diameter, and made of the best 24 gauge galvanized iron. *Smoke-Pipe.*

The warm-air pipes to have dampers placed in them near the furnace for the purpose of regulating the supply of warm air to the different apartments. *Dampers.*

The vertical warm-air pipes to be placed in walls where shown on plan, Fig. 1. Pipes to chambers No. 1, 2, 3, 4, on second floor, and to smoking and to children's room, attic, to be 3½ x 12 inch. The exposed woodwork in the walls to be carefully covered with tin. Pipes to bathroom, second floor, and bedroom in attic, 3½ x 10 inch. *Vertical Air-Pipes.*

Strong sheet-iron laths to be used to cover all pipes in walls. The joints of warm-air pipes to be soldered and pipes properly secured with strong wire attached to hooks screwed into the joists. *Laths.*

All turns in the pipes to be round, and joints seamed. *Elbows.*
Collars in register boxes to be double seamed to box. *Collars.*
Register box to fit inside of register border. *Register Boxes.*

The registers on the first floor to be black japanned, glide movement, and furnished with a cast-iron border. *Quality of Registers.*

Registers in upper rooms to be side wall, black japanned, with vertical wheel movement.

Hall, 1 14x22 register face for cold air. *First-Floor Registers.*
Hall, 1 9x14 register and border for warm air.

Parlor, 1 9x14 register and border for warm air.
Library, 1 9x14 register and border for warm air.
Dining-room, 1 9x14 register and border for warm air.

Second Floor Registers.
No. 1 chamber, 1 8x12 side wall register for warm air.
No. 2 chamber, 1 8x12 side wall register for warm air.
No. 3 chamber, 1 8x12 side wall register for warm air.
No. 4 chamber, 1 8x12 side wall register for warm air.
Bathroom, 1 10x10 side wall register for warm air.

Attic Registers.
Smoking-room, 1 8x12 side wall register, warm air.
Playroom, 1 8x12 side wall register, warm air.
Chamber, 1 8x12 side wall register, warm air.
Hall, 1 8x12 side wall register, warm air.

Registers for Ventilation.
Kitchen, 1 8x12 ventilating register.
Second floor, No. 4 chamber, 1 8x10 ventilating register.
Attic, smoking-room, 1 8x10 ventilating register, near ceiling.
Attic, smoking-room, 1 8x10 ventilating register, near floor.
Attic, playroom, 1 8x10 ventilating register, near floor.
Attic, chamber, 1 9-in. round ventilating register, near floor.

STEAM SPECIFICATION—A LOW-PRESSURE SYSTEM.

Boiler.
The steam boiler to be vertical and of wrought steel plate. Sides to be $\frac{3}{10}$ inch thick, and ends $\frac{1}{4}$ inch thick, 21 inches in diameter and 26 inches high, with 36 2-inch vertical tubes and suspended inside the dome of the warm-air furnace.

Fittings.
This combination furnace to be supplied with steam and water gauges, try-cocks, safety valve and diaphragm regulator.

Radiators.
Vertical cast-iron radiators with cast-iron fretwork tops of the following dimensions to be used in the different apartments:

First Floor.
Hall, 1 radiator, 36 inches high, containing 40 square feet   40
Parlor, 1 radiator, 36 inches high, containing 32 square feet.. 32
Library, 1 radiator, 36 inches high, containing 32 square feet.. 32
Dining-room, 1 radiator, 36 inches high, containing 32 square feet... 32
Kitchen, 1 radiator, 36 inches high, containing 44 square feet....... 44
Toilet, 1 wall coil, 5 square feet......................... ...   5

Second Floor.
No. 1 chamber, 1 radiator, 36 inches high, containing 24 square feet.... 24
No. 2 chamber, 1 radiator, 36 inches high, containing 24 square feet.. 24
No. 3 chamber, 1 radiator, 36 inches high, containing 20 square feet.. 20

No. 4 chamber, 1 radiator, 36 inches high, containing 20 square feet.. 20

Alcove, 1 radiator, 24 inches high, containing 10 square feet..:...... 10

Sewing-room, 1 radiator, 24 inches high, containing 10 square feet.. 10

Second Floor.

293

Estimate for Warm-Air Heating.

1 Combination warm air and steam furnace..........		$240.00
28 feet of 9-inch warm-air pipe (tin) ....@ 18c.		5.04
39 feet of 8-inch warm-air pipe...................... 17c.		6.63
10 feet of 7-inch warm-air pipe...................... 15c.		1.50
6 feet of galvanized iron smoke-pipe, 9 inches diam.... 40c.		2.40
4 9-inch round elbows............................. 37c.		1.48
2 9-inch bevel elbows................................ 35c.		.70
1 8-inch round elbow............................. 35c.		.37
3 8-inch bevel elbows............................. 35c.		1.05
1 7-inch bevel elbow................................ 32c.		.32
1 9-inch bevel elbow, galvanized iron, for smoke-pipe.. 40c.		.40
1 9-inch round elbow, galvanized iron, for smoke-pipe.. 50c.		.50
65 feet of 3½ x 12 tin square pipe.................... 25c.		16.25
40 feet of 3½ x 10 tin square pipe .... 22c.		8.80
4 9 x 14 tin register boxes for floor................ 50c.		2.00
4 8 x 12 tin register boxes for wall................ 40c.		1.60
2 8 x 12 tin register boxes for wall, double heads..... 60c.		1.20
1 8 x 10 tin register box for wall................... 35c.		.35
4 9-inch furnace collars............................. 20c.		.80
5 8-inch furnace collars............................. 20c.		1.00
2 7-inch furnace collars............................. 20c.		40
4 9-inch dampers for warm-air pipes................ 20c.		.80
5 8-inch dampers for warm-air pipes................ 20c.		1.00
2 7-inch dampers for warm-air pipes................ 20c.		.40
5 8-inch bevel collars for square pipe................ 25c.		1.25
2 7-inch bevel collars for square pipe............... 25c.		.50
4 9-inch ceiling rings..... .... 20c.		.80
2 9-inch thimbles for brick walls.................... 10c.		.20
2 8-inch thimbles for brick walls................... 10c.		.20
1 14 x 22 register face............................. $1.95		1.95
4 9 x 14 registers and borders....................... 3.08		12.32
8 8 x 12 side wall registers......................... 1.40		11 20
1 8 x 10 side wall register...........................		1.25
1 9-inch round ventilator...........................		1.00
4 8 x 10 ventilators.................................@ 1.25		5.00
1 8 x 12 ventilator.................................		1.40
8 feet of galvanized pipe for ventilating chamber attic..@ 30c.		2.40
10 sheets of 20 x 28 I. C. tin plate for tinning exposed woodwork... 15c.		1.50
1 tinner four days' work fitting in pipes............. $3.00		12.00
1 helper, four days' work fitting in pipes........ .... 2.00		8.00
Total		$355.96

Estimate for Steam Heating.

122	feet of 2-inch steam-pipe	@ 20c.	$24.40
10	feet of 1½-inch steam-pipe	14c.	1.40
185	feet of 1¼-inch steam-pipe	11c.	20.35
25	feet of 1-inch steam-pipe	9c.	2 25
60	feet of ¾-inch steam-pipe	6c.	3.60
35	feet of ½-inch steam-pipe	4c.	1.40
6	2-inch elbows	20c.	1.20
4	2 x ¾ inch elbows	23c.	.92
3	1½-inch elbows	12c.	.33
40	1¼-inch elbows	10c.	4.00
4	1¼ x ¼ inch elbows	12c.	.48
4	1-inch elbows	6c.	.24
12	¾-inch elbows	5c.	.60
4	½-inch elbows	3c.	.12
2	2-inch T	30c.	.60
11	2 x 2 x 1¼ T	35c.	3.85
1	1½ x 1¼ x 1¼ T	....	.22
1	1½ x 1½ x 1¼ T	....	.22
1	1¼ T	....	.15
1	1¼ x ¾ T	....	.18
2	2-inch flange unions	@ 75c.	1.50
17	1¼-inch plated floor-plates	15c.	2.55
1	1-inch plated floor-plate	15c.	.15
1	¾-inch plated floor-plate	15c.	.15
5	1¼-inch plated ceiling-plates	15c.	.90
6	2-inch pipe-hangers	35c.	2.10
1	1-inch pipe-hanger	....	.15
25	pounds asbestos	@ 23c.	5.75
165	square feet hair felt	8c.	13.20
26	square yards of canvas	25c.	6.50
293	square feet of radiating surface in 12 radiators	at 45c. per square foot.	131.85
11	1¼-inch plated valves	@ $2.25	24.75
1	1-inch plated valve	2.00	2.00
1	¾-inch plated valve	1.75	1.75
	(Brass cocks for blow-off pipe, water-feed pipe and return-pipe are included in price of boiler.)		
12	plated automatic air-valves	@ $1.25	15.00
	Bronze for radiators and pipes	....	6.00
1	patent boiler flue-cleaner	....	8.00
1	steam fitter, 10 days at $4.00	....	40.00
1	helper, 10 days at $2.50	....	25.00
	Total		$353.84

HOT WATER AND HOT AIR.*

BY A. R. BRINK.

ADVANTAGES AND GENERAL FEATURES.

Accompanying this are plans for heating the competitive house by hot-water and hot-air combination. I made the floor plans of each floor broken, merely to show where the radiators and registers are to be located. By the plan I send you of heating I think you will find it superior to many others. I take the warmed air from furnace into the rooms with a tin pipe, which leads to the back of radiators and forces the heat of radiator out into the room, and gives a supply of fresh air without taking it direct from the outside of the house. Neither will it cool the radiator off so quickly as if it were drawn from outside when the fire is low. I suppose there will be some objection to having the radiators placed against the inside walls, but with the air coming in at the back it will drive the heat out into the room, while if they were placed on the outside walls without the air back of them, there would be no circulation. I have this style of arrangement in my own house, and find it just as warm in all parts of the room as it is 6 feet from the radiator, and one of the radiators is placed on the south wall inside of a northwest room exposed on two sides.

You will see by the drawings how the pipes run and the size of them, the location of radiators; also the amount of square feet each one contains. The hot-air pipes back of radiators I have made very small—6 inches in diameter. It is not necessary to have them larger, for the heat of the radiators will have a tendency to create an upward circulation which will furnish all the fresh air that is necessary to a good ventilation.

SPECIFICATION.

Pipe A, Fig. 1, in south partition leads to southwest chamber second floor, Fig. 3, which cannot come back of radiator and work well on account of the distance; it is merely to furnish fresh warmed air to this room. Size of this partition-pipe, 3 x 8 inches square; round pipe 6 inches. Pipe B leads to library and dining room, Fig. 2; size, 8 inches round, 3 x 10 inches square, to be made as shown in detail, Fig. 7. Pipe D leads to billiard-room and chamber third floor, Fig. 5, 10 inches round, and 3 x 14-inch partition-pipes.

(margin note: Hot-Air Pipes in Cellar and Partition.)

* From *The Metal Worker*, April 6, 1889. Copyrighted, 1889, by David Williams.

Pipe C to back of parlor radiator, Fig. 2, 6 inches round and 3 x 8-inch partition Pipe E into chimney flue to playroom, Fig. 5, 9 inches round. Pipe F to back of hall and toilet, Fig. 2. Pipes F, E, each lead to front rooms second floor, Fig. 3, to carry hot air from coil in box. Size of partition-pipe, 4 x 12 inches. · Box and outlet to be made as shown on drawing, Fig. 4. Box to be made large enough to receive a coil of pipe which contains 75 square feet of heating surface. This box to be made of bright tin with cold-air inlet at bottom, as shown. Cold air inlet to have a damper, also the upright hot-air pipes.

Smoke Pipe. Smoke-pipe to be made of No. 18 iron, riveted together, and to have a Lawson check-draft damper. This pipe to be made so that it can be taken down (without breaking) when necessary to clean it out. **Hot-Air Pipes.** All hot-air pipes to be made of bright IC tin, double seamed, and each is to have a damper close to the furnace that will remain where set. All partition-pipes to be lined outside with tin (or asbestos felt) ½ inch from any wood. Pipes to be connected to heater, as shown in Fig. 1, and to have as much slope toward outlet as can be given them. No sharp bends allowed. Wall registers to be fastened onto upright partition-pipes with iron frames **Wall Registers.** riveted to openings. Wall registers all to be japanned convex registers. Sizes of all shown on plans.

Ventilation First and second floors to be ventilated by fireplaces. Billiard-room into chimney of same room. Playroom to have a square pipe made of roofing tin, placed in partition where shown, Fig. 5, and run overhead into furnace flue vent. Bathroom, Fig. 3, where shown, and run overhead into kitchen vent-flue. Toilet vent to run under first-floor joist into furnace flue vent as shown, Fig. 1.

Heater. The heater is to be located and bricked in with a 4-inch wall, and covered with T iron and two courses of brick. Cold-**Cold-Air Duct.** air duct to be built in, size as shown, Fig. 1, under cellar floor to furnace, and up to window on outside wall of building. Cold-air duct up to furnace from hall, Fig. 2, to be made of roofing tin 8 x 24 inches, and run under first-floor joist to furnace. Both ducts to have a damper that will remain where set. Cold-air opening in hall to be covered with a 12 x 24 register face.

Hot-Water Circulation. All flow and return pipes to be in size as marked on the plans, and run where shown, Figs. 1 and 4. No right angle **Fittings.** to be used unless absolutely necessary. Use 45° Ls and Ys where bend or branch is to be made in the run. Flow and return pipes to have a rise from heater not less than ¼ inch **Expansion Tank.** in each foot in length. Expansion tank to be made of No. 20 galvanized iron, riveted and soldered tight, to be 48 inches long by 14 inches wide; to have glass water gauge, and 1-inch

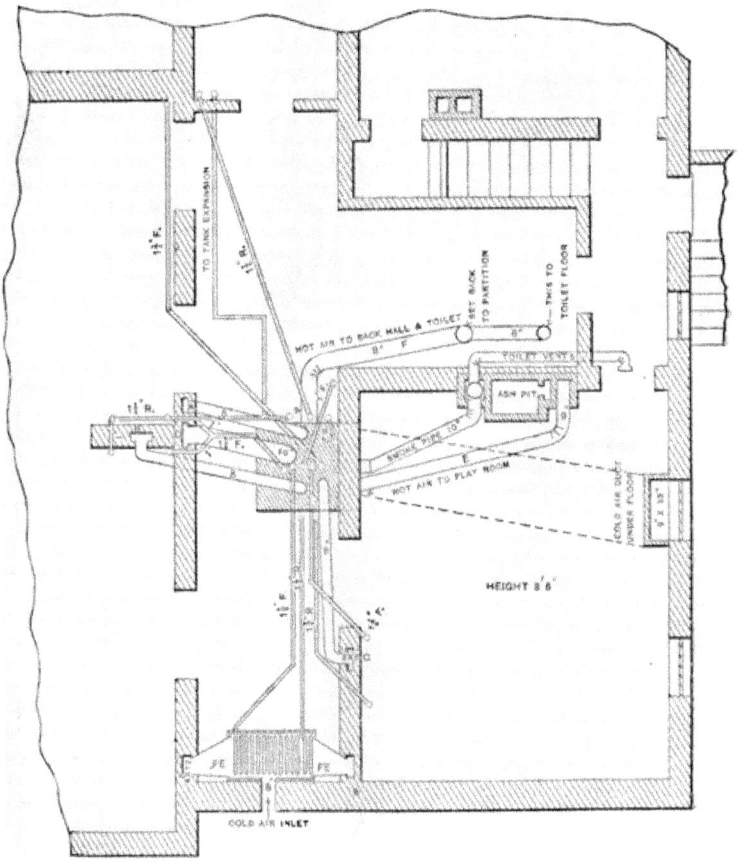

Fig. 1.—Cellar Plan.

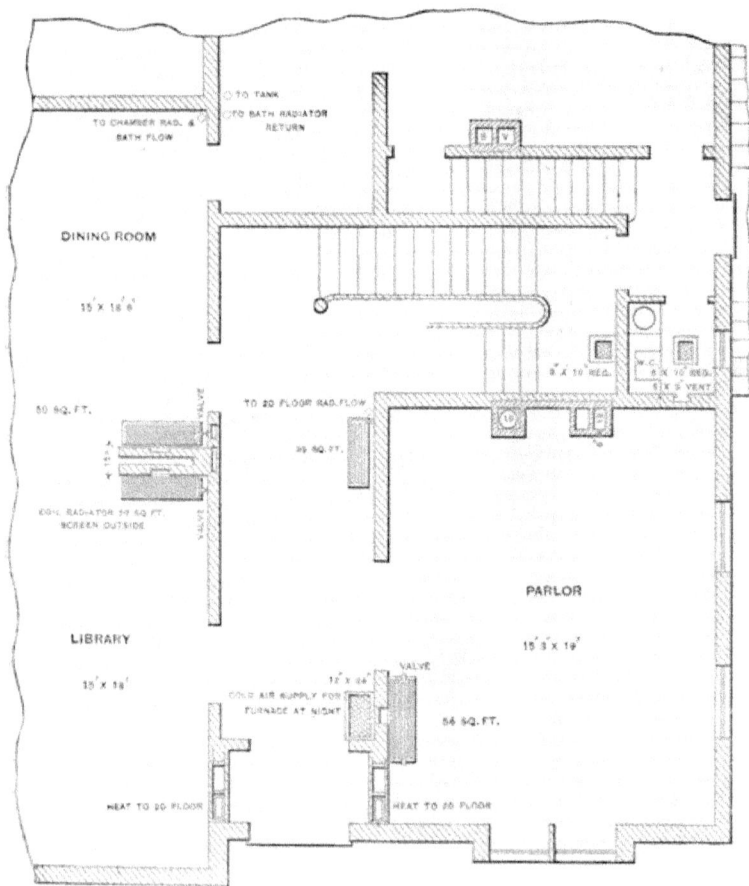

Fig. 2.—First-Floor Plan.

pipe from top out through roof. Make as shown in detail, Fig. 6.

Run of pipes and size can be seen in Figs. 1 and 4. Floor plans will show the location of the radiators and number of square feet in each. Valves to be placed on radiators as marked. All valves to be slide-valves, or any kind that will open full and not obstruct the circulation. Radiators all to have an air-cock at highest point on radiator. Place a draw-off valve at lowest point on heater. Provide a force pump in cellar to fill tank, with pipe leading from it to cistern, as shown in Fig. 4. All radiators on first floor and hall second floor to be covered with ornamental screens; if of iron, they are to be bronzed. *(Size of Pipes. Size of Radiators. Valves. Water Supply. Radiator Screens.)*

Radiators to be made of 1-inch pipe, and return bends with 1¼-inch inlets and outlets. Bends to be tapped on a slight bevel, in order to give all pipes a downward slope. *(Construction of Radiators.)*

Place a valve on pipe at pump. Heater to have at least 40 square feet of heating surface exposed to the fire. Heater to be bricked in as indicated in Figs. 1 and 4. All of the above to be done in a good, workmanlike manner, and true to the specifications. *(Boiler-Heating Surface. Brick Setting.)*

I am not a manufacturer of heaters, but handle any kind that is good. Of course any style of heater can be used for heating the house by the arrangement I send you, if it has heating capacity for such a house. The heater which I specify is one that I invented about two years ago, and intended to have it patented, but I find the market is flooded with all kinds of heaters, so have given up the idea of spending any more money upon it. It is a base-burning hot-air heater, as well as a powerful hot-water heater. The whole of the water space is surrounded by the fire. The hot-air part of it is to be perfectly air-tight, so that no gas or smoke can escape into the hot-air chamber. *(Heater.)*

The heater has a good-sized feed fire door and a large ash-pit, also a shaker and clinker-grinding grate, which can be drawn out very easily when it is necessary to dump the fire. I will state that it works something on the principle of some grates that are now in use, but I do not think there is any infringement upon them. At least I have tried not to make it so.

Fig. 8 shows how the two radiators are to connect with one flow and return pipe. *(Radiator Connections.)*

There are a few changes to be made in the building, such as making the partition between library and dining-room 15 inches instead of 12 inches. *(Changes in House.)*

The partition on each side of front vestibule should be made of 6-inch studding, as the hot-air pipe is 4 x 12 inches and will require that.

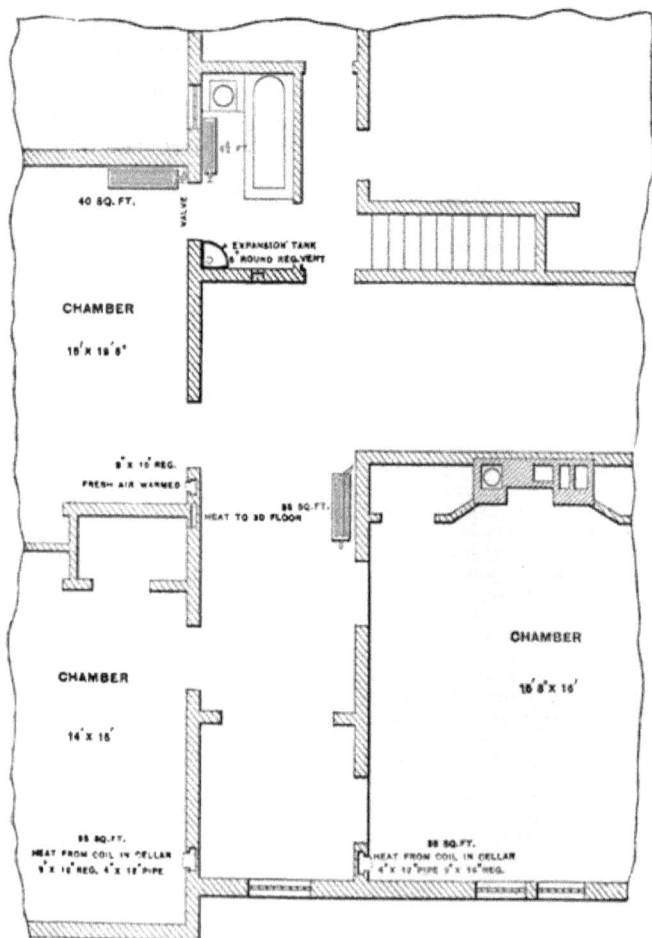

Fig. 3.—Second-Floor Plan.—Scale, ⅛ Inch to the Foot.

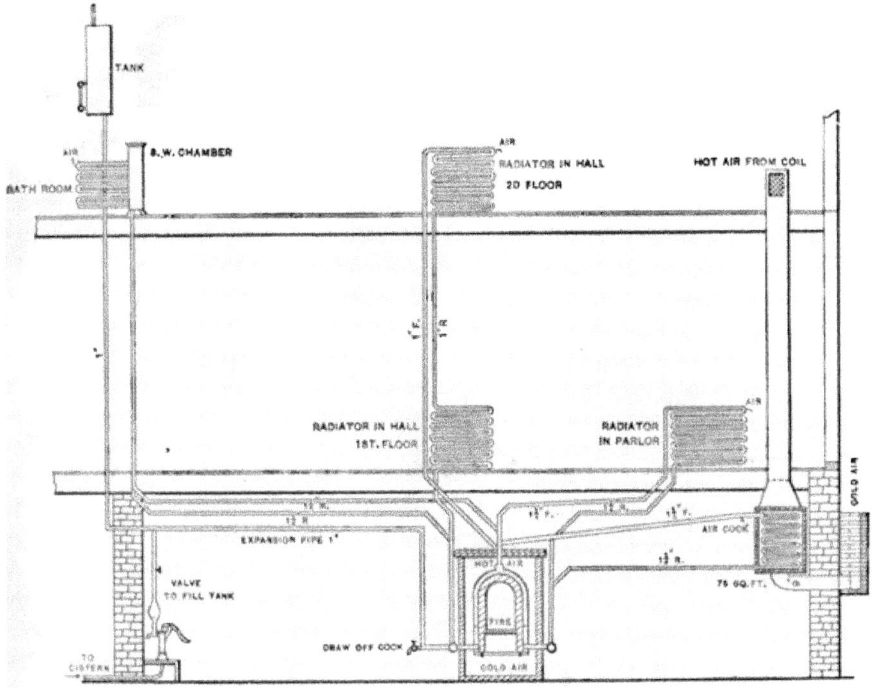

TANK

S. W. CHAMBER

AIR

BATH ROOM

AIR
RADIATOR IN HALL
2D FLOOR

HOT AIR FROM COIL

RADIATOR IN HALL
1ST. FLOOR

RADIATOR
IN PARLOR

AIR

COLD AIR

EXPANSION PIPE 1"

VALVE
TO FILL TANK

AIR COCK

75 SQ. FT.

DRAW OFF COCK

HOT AIR

FIRE

TO
CISTERN

COLD AIR

Fig. 4.—Sectional Elevation.

The chimney on north side of house should be built as I
have drawn it, with a 10-inch clay pipe built inside (space
outside of to be used for ventilator) of brickwork for furnace
flue and a hot-air flue to carry heat to playroom. The clay
pipe in the chimney should run at least 2 inches above top of
chimney. The expansion tank, Fig. 6, I have shown much
larger than is necessary to hold the water required, but I
make it large so that it will not waste when the heater is
running hard. There should be about 1 foot of water in the
tank when the heater is working easy. The third floor,
Fig. 5, is entirely warmed by hot air. There will be times
when no heat is required on this floor, and if hot water is
used it is liable to freeze if turned off on a cold day. Hot
air will flow very well to this floor, as it is high and a good
draft for it.

ESTIMATE.

Itemized account of labor and material of hot-air and hot-water combination of
A. R. Brink.

HOT AIR.

6½ feet I. C. bright tin 10-inch pipe ....@ 30c.		$1.95
10 feet I. C. bright tin 9-inch pipe...................	25c.	2.50
18 feet I. C. bright tin 8-inch pipe...................	22c.	3.96
17 feet I. C. bright tin 6-inch pipe......	15c.	2.55
25 feet I. C. bright tin 3 x 14 pipe...................	35c.	8.75
29 feet I. C. bright tin 4 x 12 pipe...............	35c.	10.15
25 feet I. C. bright tin 3 x 8 pipe...................	25c.	• 6.25
1 9-inch elbow I. C. bright tin...........		.75
3 8-inch elbows I. C. bright tin.....................@ 70c.		2.10
1 10-inch elbow.............................		.80
4 6-inch elbows.................................@ 50c.		2.00
50 sheets 20 x 20 roofing tin, lining............ ...	15c.	7.50
Vent-pipe from toilet.		2.50
9 feet 10-inch No. 18 iron........................@ 40c.		3.60
1 Lawson damper.......		1.00
8 hot-air pipe dampers........		1.80
Tin box for coil under hall.........................		8.00
8 feet cold-air duct 8 x 24 roofing tin...............		2.80
Vent-pipes, playroom and bathroom.................		10.00

REGISTERS.

4 8 x 10 side-wall convex registers.................@ $2.00		8.00
4 8 x 12 side-wall convex registers.................	2.25	9.00
1 6 x 8 side-wall convex register..............		1.75
1 9 x 12 side-wall convex register..................		3.00
1 6-inch round register............		1.00
1 8 x 10 register and border (floor).................		2.00
1 6 x 10 register and border (floor).................		1.75
1 12 x 24 register face (hall, cold air).............		1.75
		——— $107.21

CHAMBER

HOT AIR 8 X 18 REG.

HOT AIR 8 X 18 REG.

HOT AIR 9 X 18 REG.

PLAY ROOM

VENT PIPE TO RUN
OVER HEAD TO FURNACE
FLUE 5 X 18 PIPE
8 X 18 REGISTER

BILLIARD ROOM

VENTILATE THIS ROOM IN THE
CHIMNEY ON SOUTH SIDE WITH
8 X 18 REGISTER AT BASE

ONE INCH PIPE

OUT THROUGH ROOF

14

48

WATER LINE
GAGE

1 PIPE

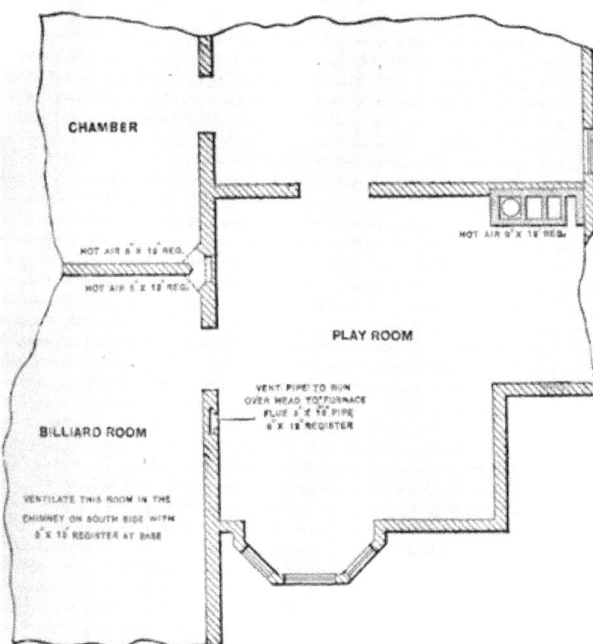

Fig. 5.—Attic Plan.   *Fig. 6.—Expansion Tank.*

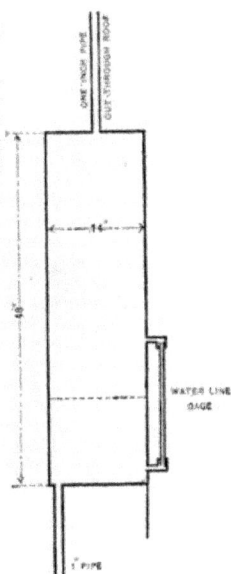

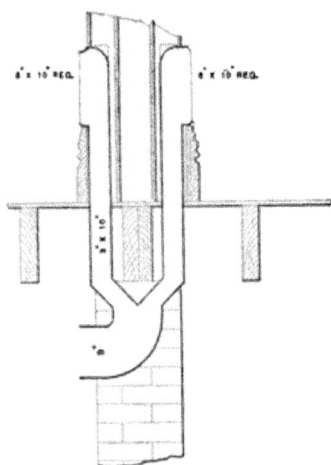

Fig. 7.—*Hot Air Pipe to Library and*
Dining-Room.

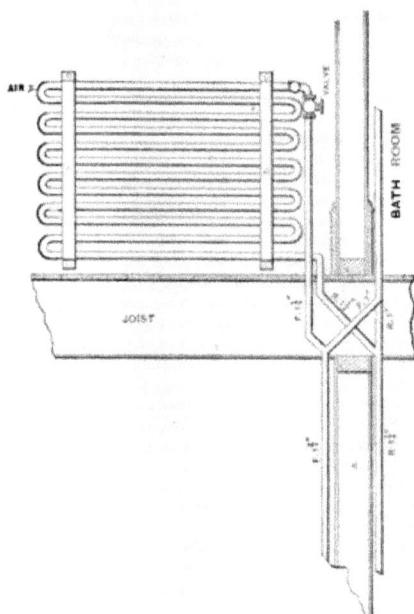

Fig. 8.—*Radiator Connections.*

Brought forward.... . . .. $107.21

1138 brick in heater walls......................@ $7 M	$7.97	
500 brick in cold-air duct......................... "	3.50	
Stone to cover cold-air duct	5.00	
Time laying brick, with lime and sand..............	15.00	
Work and material on cold-air inlet to coil..........	3.00	
Dampers in cold-air ducts.........................	1.25	
Time setting pipes, cutting holes and register work..	15.00	
		50.72

Hot-air attachments. $157.93

Cost of Piping and Radiators:

132 feet 1¼-inch pipe .... @ 11c.	$14.52	
100 feet 1-inch pipe.. .... 8c.	8.00	
24 1¼-inch elbows, 45° .... 20c.	4.80	
10 1-inch elbows, 45° .... 15c.	1.50	
2 1¼-inch Ys..... .... 40c.	.80	
15 1-inch elbows... .... 12c.	1.80	
6 1-inch Ts...... .... 15c.	.90	
4 1¼-inch slide-valves, nickel...................... $2.75	11.00	
1 1-inch valve to pump pipe........................	1.50	
1 1-inch valve to draw-off pipe....................	1.50	
1 expansion tank, with gauge......................	6.00	
2 radiators of 50 square feet each....	50.00	
2 radiators of 35 square feet each.................	35.00	
1 radiator of 56 square feet.......................	28.00	
1 radiator of 40 square feet.......................	20.50	
1 radiator of 7 square feet.......................	3.50	
1 box-coil under hall, 75 square feet...............	37.50	
70 square feet of ornamental screen for radiator......	35.00	
Radiators, hangers and stands, &c..................	15.00	
Force-pump....	8.00	
Bronze and japan.	4.50	
7 air-valves..... ....@ 15c.	1.05	
Excavating for furnace and air duct. &c.............	10.00	
Steam-fitter's and helpers' time	58.00	
		358.37
Combination heater ready to set in brick-work........		255.00
Cost of whole job complete...................		$771.30

II. STEAM CIRCULATION.

STEAM CIRCULATION.*

BY E. P. WAGGONER.

SPECIFICATIONS.

For the labor and materials required in the erection and entire completion of a low temperature return circulation, steam warming and ventilating apparatus, in full accordance with the following specifications.

The contractor is to give his personal supervision to the work, to furnish all transportation, labor, materials, apparatus, scaffolding and utensils needful for performing the work in best manner, according to the specifications. Should the contractor introduce any materials different from the sort and quality herein described or meant to be implied, it shall be immediately removed at contractor's expense any time during the progress of the work. The works are to be executed in the best, most substantial and thorough workmanlike manner according to the true intent of these particulars, and which are intended to include everything requisite and necessary to the proper and entire finishing of the work, notwithstanding any item necessarily involved in the work is not particularly mentioned, and all, when finished, to be delivered up in a perfect and undamaged state. *General Condition.*

Furnish and set complete, where shown on plan, Fig 1, one upright, return flue, magazine feed, portable boiler, cast iron, sectional. *Boiler.*

The boiler to be set on a good solid foundation, made of 4 inches hard-burned brick laid in cement mortar; foundation to be 6 feet wide and 8 feet long. *Boiler Foundation.*

The boiler to have not less than 140 square feet of heating surface. *Heating Surface.*

The boiler to have not less than 4¾ square feet of grate surface. *Grate Surface*

The boiler will be provided with the following trimmings: Safety-valve, nickel-plated steam-gauge, water-column, gauge-cocks, water-gauge, blow-off valve and automatic draft regulator. *Boiler Trimmings.*

The boiler will be provided with the following tools: Wrought-iron poker, hoe and flue brush. *Tools.*

---

* From *The Metal Worker*, May 4, 1889. Copyrighted, 1889, by David Williams.

Automatic Water-Feeder. Furnish and set complete one improved automatic water-feeder, connected to the return and steam mains ; also with the water-pipe from tank. Place in the water and steam pipes and discharge-pipe from feeder asbestos-seat globe-valve.

Steam-Supply Pipes. The main steam-supply pipes are to be run as shown on the plan, Fig. 1. The main pipe will be 2½ inches internal diameter. There will be two branches from the boiler, to supply steam to the various points of radiation. These pipes will be carefully graded to a pitch from the boiler of ¼ inch in 10 feet and securely hung in position by wrought-iron hangers. The arrangement to be such as to provide ample compensation for expansion.

Return-Pipes. The return-pipes are to be run as shown on the plan, Fig. 1, and of the size marked. All the main return to be below the water-line in boiler. At connection to boiler place a valve to drain the returns, and connect pipe to drains. All the return-pipes shall have a carefully-graded pitch toward the boiler.

Drips. At all points where condensed steam can by any means collect in the steam-pipes there will be located drip-pipes not less than 1 inch, connecting the steam to the return lines.

Rising Pipes. There shall be the number of rising pipes shown on plan, Fig. 1, to connect to radiators on first and second floors. They will be run as shown on plan, and of the size marked.

All the branches from main steam-pipe to be taken from the top, and all branches to drip back into the main. All branch pipes to be one size larger than the rising pipe.

Return Risers. All the return-pipes from radiators to be run as shown on plan, Fig. 1, and of the size marked. All to connect to the main return below the water-line of boiler.

Check-Valves. Each return-pipe from the radiators to run 1 foot horizontal in the cellar, above the water-line in boiler. Place in each return at this point horizontal swing check-valves; also place one swing check-valve 2½ inches at connection of returns to boiler.

Protectors. At all points where steam or return pipes pass through ceilings, floors and partitions the channels and holes are to be protected or lined with metal tubes and cast-iron floor and ceiling plates. Each connector to radiators to have nickel-plated carpet plates.

Pipe Covering. All steam, return and drip pipes from air valves to second floor to be covered with 1/16-inch asbestos board, ¾-inch hair felt and 10-ounce canvas, all secured with copper wire to pipe. All steam and return pipes in room under parlor to be covered same as risers.

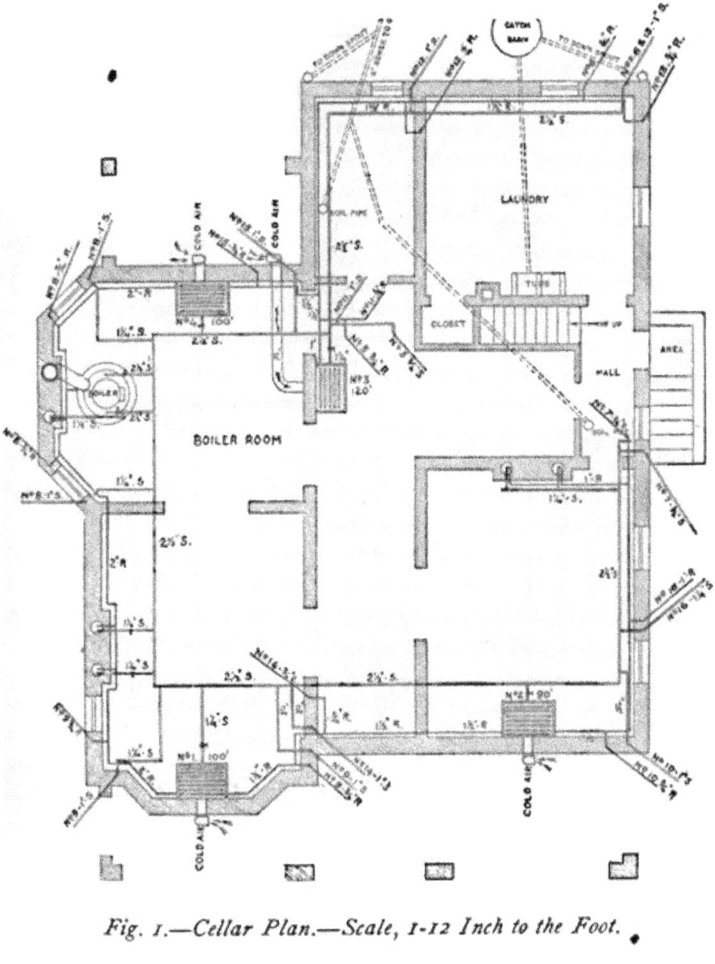

Fig. 1.—Cellar Plan.—Scale, 1-12 Inch to the Foot.

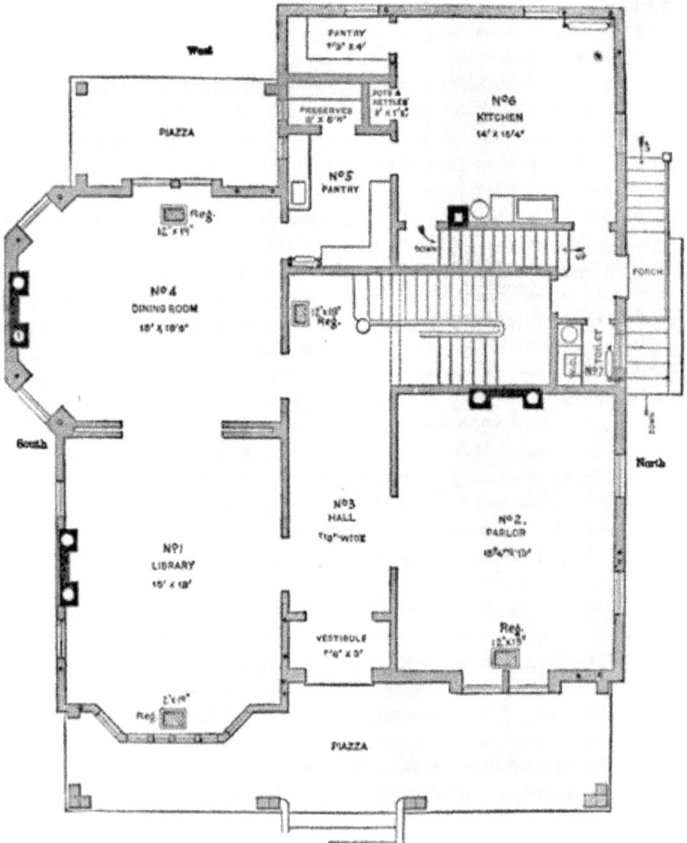

Fig. 2.—First-Floor Plan.—Scale, 1-12 Inch to the Foot.

Each radiator will be provided with two radiator-valves, nickel-plated, all-over finished trimmings, with wood wheel and asbestos seats. The connection between valve and radiator shall be by a ground-joint union. **Radiator Valves.**

Each radiator and each indirect stack and on end of each main steam-pipe to have nickel-plated automatic air-valves with $\frac{1}{8}$-inch drip-pipes connected to $\frac{1}{2}$-inch pipe in cellar, the $\frac{1}{2}$-inch connected to drain. **Air Vents.**

Furnish and set complete two electric heat regulators. One to be connected to damper in smoke-pipe, draft door under grate and mixing-valves on indirect radiators to hall and parlor. The other to be connected to mixing-valves on indirect radiators to library and dining-room. All to be set complete with all wire cable, pulleys, wiring and batteries. One thermostat to be located in hall and one in library. **Heat Regulators.**

The system of piping shall be so arranged as to secure a complete and perfect circulation of steam throughout the building without any pressure indicated on the steam-gauge at the boiler; and there shall be no noise in the pipes by reason of condensed steam or expansion. **Circulation.**

Furnish and set the following vertical-tube radiators: **Radiators, Direct.**

Pantry, one,	12	feet	of heating	surface,	valves	........$\frac{3}{4}$ x $\frac{3}{4}$
Kitchen, one,	36	"	"	"	"	........1 x $\frac{3}{4}$
Toilet, one,	5	"	"	"	"	........$\frac{3}{4}$ x $\frac{3}{4}$
Chamber, two,	42	"	"	"	"	........1 x $\frac{3}{4}$
Chamber, one,	45	"	"	"	"	........1 x $\frac{3}{4}$
Chamber, two,	40	"	"	"	"	........1 x $\frac{3}{4}$
Chamber, one,	30	"	"	"	"	........1 x $\frac{3}{4}$
Sewing, one,	10	"	"	"	"	........1 x $\frac{3}{4}$
Bath, one,	14	"	"	"	"	........1 x $\frac{3}{4}$
Billiard, one,	30	"	"	"	"	........1 x $\frac{3}{4}$
Chamber, one,	20	"	"	"	"	........1 x $\frac{3}{4}$
Play-room, one,	60	"	"	"	"	.......$1\frac{1}{4}$ x 1

All the radiators are to be bronzed with the best gold bronze. All the pipe and all cast-iron work on boiler to be painted one coat of maroon japan.

Furnish and set complete the following indirect radiators: **Indirect Radiators.**

Parlor, one stack, 90 feet of heating surface.
Hall, one " 120 " " "
Library, one " 100 " " "
Dining-room, one stack, 100 feet heating surface.

The indirect stacks will be securely suspended from cellar ceiling by wrought-iron hangers, to be ceiled overhead and incased in matched and beaded pine ceiling lined with bright tin. All the stack boxes to be put together with screws. A **Stack-Boxes.**

space 12 inches high to be left above and below the indirect radiators.

Hot-Air Ducts. Hot-air ducts are to be run to registers located in rooms above. No hot-air duct to be less than 12 x 19 inches.

Cold Air to Radiators on Second Floor. Galvanized-iron cold-air pipes 4 x 12 inches are to be run from the ceiling of piazza through the building and connect to 4 x 12 register under radiators in rooms Nos. 9 and 10 (Fig. 3); register faces to be placed over pipes on ceiling of piazza.

Cold-Air Ducts, Tile. The cold air to the indirect radiators will be taken from the outside of the building, as shown on plan, Fig. 1, through vitrified tile. One tile to go through the foundation wall with elbow on end, and one up to surface of the ground with elbow on top, the elbow on top to have fine wire screen on iron frame.

Cold-Air Ducts, Galvanized Iron. From the tile in foundation wall connect to indirect radiators, with galvanized-iron pipes, No. 24 iron, well riveted and soldered, with flange on end to screw to mixing boxes on indirect stack boxes.

Mixing-Boxes. On outside of indirect boxes square mixing-boxes are to be built of the size of cold-air supply, with openings above and below the indirect radiators full size of cold-air pipe.

Mixing-Valves. In each mixing-box place wrought-iron damper to connect with lever outside, so that the air can be turned through the radiators or over them. Electric regulators to be connected to this damper.

Cold-Air Dampers. In each cold-air supply to indirect radiators place a wrought-iron damper, to regulate air to the radiators.

Registers. Furnish and set in place (Fig. 2) four 12 x 19 nickel-plated end-wheel registers with valves.

Smoke-Pipe. Connect the boiler with the smoke-flue with galvanized-iron smoke-pipe made of No. 20 iron, well riveted together Smoke-pipe to have a wrought-iron damper, with 3-inch round hole through it in center, with lever to connect electric heat regulator.

Valves on Indirect Radiators. Each stack of indirect radiators to have $1\frac{1}{4}$-inch steam and 1-inch return valves. Valve to be asbestos seat, brass globe valves, iron wheel.

Ventilation. Rooms Nos. 1, 2, 4, 8, 9, 10 (Figs. 2 and 3) to have 8-inch vitrified tile built into the chimneys from the cellar to top of the chimneys, as shown on plan, Fig. 1, with openings into the fire-place of rooms they connect with. At the bottom of each flue in cellar two openings are to be left, one to connect the heating-pipe and one to clean out flue.

Pipe in Vent-Flues. In each vent-flue run two lines of $1\frac{1}{4}$-inch pipe up to the floor of the attic, one pipe to be connected to the main steam-pipe and one to the main return-pipe. Each set of pipes to have $1\frac{1}{4}$-inch globe-valve on steam-pipe and $1\frac{1}{4}$-inch swing-

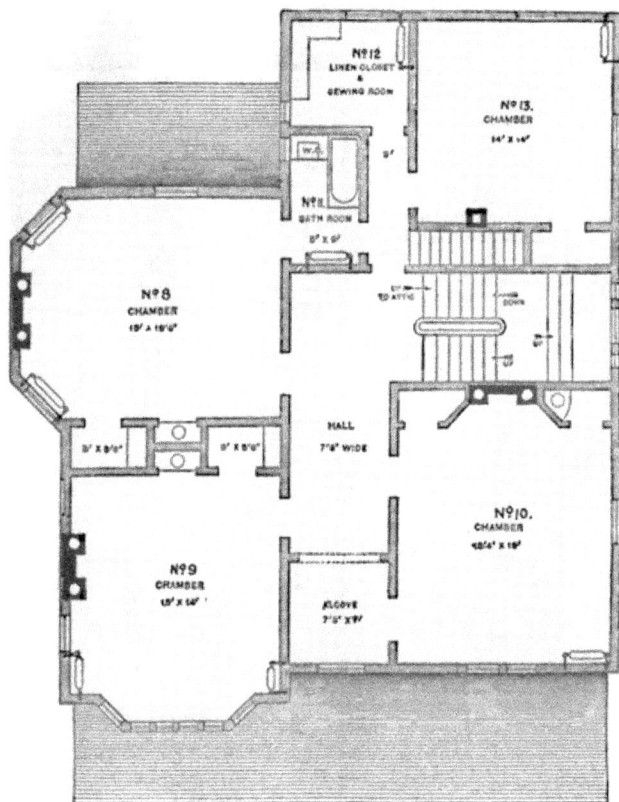

Fig. 3.—Second-Floor Plan.—Scale, 1-12 Inch to the Foot.

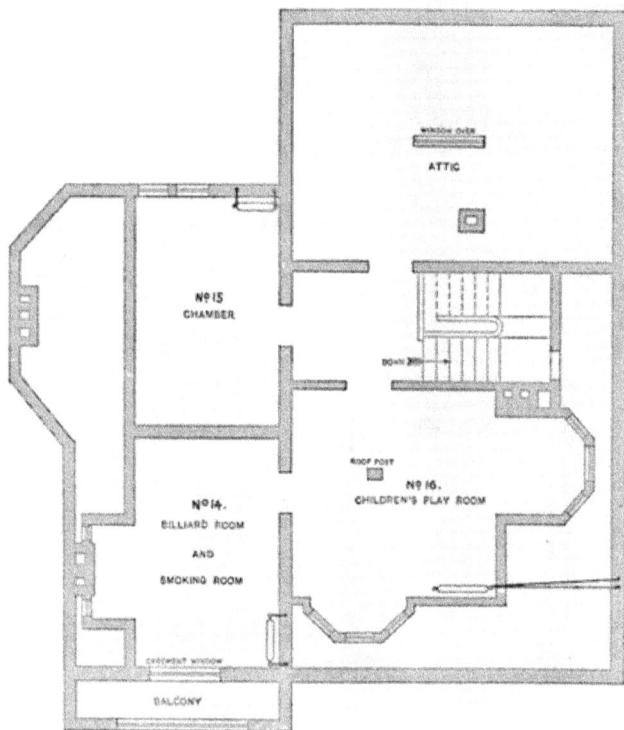

Fig. 4.—Attic Plan.—Scale, 1-12 Inch to the Foot.

check on return in cellar. Each return to have automatic air-valve in the cellar. All openings into tile in cellar to be made air-tight.

The smoke-flue for boiler to be 12-inch vitrified tile, built into the chimney, with openings into the cellar for smoke-pipe and cleaning door (Owner to furnish all the tile-flues.) Tile Smoke-Pipe.

We propose to erect complete the low temperature steam-warming and ventilating apparatus as described by this specification, and subject to the following guarantee: Proposal.

First, To warm all rooms in any weather with not to exceed 2 pounds pressure at the boiler.

Second, To warm and ventilate all rooms that the apparatus is connected with to a temperature of 70° F. in any weather.

Third, That the apparatus will be noiseless and perfect in its operation.

Fourth, That the boiler will fill the entire radiation with steam for ten hours without attention.

ESTIMATE.

Boiler complete...	$245.00	Hot-air ducts.......	$5.00
Boiler foundation...	5.00	Cold air to second floor, galvanized-iron pipe...	4.50
Automatic water-feeder...	15.00		
Pipe and fittings...	123.50	Tile air-ducts ...	20.00
Check-valves ...	20.90	Galvanized-iron ducts...	9.45
Pipe covering...	21.25	Mixing boxes...	5.00
Radiator valves...	30.00	Mixing valves...	1.50
Air-valves ...	22.00	Cold-air dampers...	1.50
Heat regulators...	60.00	Registers...	12.50
Direct radiators...	87.10	Globe valves...	10.00
Bronzing and painting...	12.50	Smoke-pipe...	5.00
Indirect radiators ...	65.60	Labor ...	90.00
Stack boxes ...	40.00	Freight and cartage...	25.00
Hangers and tin...	12.00		
		Total ...$949.30	

ADVANTAGES AND GENERAL FEATURES.

Reasons why we specify the use of the material we have in accompanying specifications are given below:

Boiler.　　We use a cast-iron sectional boiler for the generation of steam because the same is absolutely safe under all conditions, and under the grossest mismanagement can be injured but a sectional part, which can readily be replaced at small expense.

The boiler is portable, doing away with expensive brick-work and making the same easy and convenient to handle into the house through any ordinary door and stairway without making any special provision for admitting the boiler in the bulk.

It is provided with magazine feed in order that a supply of coal sufficient to last all day or all night can be provided at one time and same be fed to the fire as necessity requires.

Boiler Foundation.　　A solid foundation laid in brick and cement is provided, that the boiler may always retain its level and upright position and not be affected by settling, as would be the case if the foundation were soft.

Heating Surface.　　We provide a boiler of 140 square feet of fire surface, allowing 129 square feet of fire surface to supply the direct and indirect radiation set in the building, while the 11 square feet of fire surface will be sufficient to heat the unprotected pipes in the building and allow the boiler to carry the job without extra effort.

Grate Surface.　　We allow 1 square foot of grate surface to 30 square feet of fire surface in the boiler.

Boiler Trimmings.　　Safety-valve is provided to avoid, under any circumstances, an overpressure of steam. Steam-gauge is provided to indicate at all times the pressure of steam on the boiler. The water-gauge is provided for the purpose of at all times showing the water-line in the boiler. The blow-off valve is provided for drawing off the water in the boiler whenever necessary. An automatic draft regulator is provided, which automatically opens and closes both the direct and the check drafts on the boiler, and by this means keeps up a steady, uniform condition of the fire.

Tools.　　The boiler is provided with a poker and hoe, for the purpose of cleaning the fire and assisting in removing the ashes. The flue-brush is for the purpose of removing the accumulated ashes with the brush out of the return-flues.

Automatic Water Feeder.　　An automatic water-feeder is provided, which supplies automatically the necessary water to the boiler and keeps a uniform quantity in same, which is very essential and desirable in

STEAM CIRCULATION. 49

generating steam. The feeder is provided with valves on both flow and feed, which may be closed at any time when it becomes necessary to repair the feeder without interfering with the running of the boiler.

Steam-pipes will be full size to the end of each main, so that all the radiation can fill without pressure. *Steam-Pipes.*

Return-pipes are below the water-line to secure water-seal, so that steam cannot fill radiators from return-pipes. These return-pipes are provided with valve near boiler, in such position that the return-pipes can all be drained of water at any time necessary. *Return-Pipes.*

Drips are provided at all points where condensed steam may collect, in order to conduct this condensation to the return-pipes in a noiseless and proper manner. *Drips.*

Branches will be taken from the top of the steam main, in order to secure at all times the dryest steam in the circulation. The branches will be one size larger than the risers, to provide ample supply for the radiation reached by them. The branches and riser-mains will drop back into the main steam-pipe, and be graded in such a manner as to avoid any noise from the passage of the condensation into the main. *Rising-Pipes.*

Return risers will connect with main return below the water-line to avoid any possibility of noise in operation. *Return Risers.*

Check-valves will be placed in the returns at points indicated, so that the supply of steam to the radiators can be regulated to heat the rooms to any temperature. *Check-Valves.*

Tin protecting tubes will be placed, as specified, in floors and partitions to protect the wood-work from coming in contact with the heating-pipe. *Protectors.*

The steam, return and drip pipes are covered as specified, for the purpose of preventing radiation of heat and the consequent condensation of steam, this materially affecting the consumption of fuel in the boiler; as the hotter the condensation can be returned to the boiler the less fuel will be required to raise this water to the steam-making point again. *Pipe Covering.*

Radiators are supplied with valves, for the purpose of closing off the heat whenever required. *Radiator Valves.*

Air-vents are placed in each radiator, whether direct or indirect, to secure perfect circulation, as such circulation cannot be secured in radiators when there is any air in them. *Air-Vents.*

The automatic heat regulators are provided to act upon the dampers of the boiler, entirely controlled by the temperature of the rooms in which the regulators are placed, thus securing a uniformity in the temperature otherwise unattainable *Heat Regulators.*

Indirect radiators are supplied for heating the first floor, with the view of properly heating and ventilating the apartments that are most occupied by assemblages, at the same *Radiators.*

time not offering any obstruction, as would be the case if
direct radiators were used on this floor, as the heating and
ventilating is accomplished by this system entirely through
registers placed in the floors or walls of the room.

Direct radiation is provided for the upper rooms of the
house to secure proper temperatures in those apartments, and
one in the kitchen to protect the water-pipes in extreme
weather.

Stack-Boxes. We line the stack boxes with tin to reflect the heat, leaving
ample space above and below the coil for the circulation of
air.

Hot-Air Ducts. Hot-air ducts will be constructed of tin and lead from the
stack boxes to the registers opening into the rooms to be
heated.

Cold-Air Ducts. Cold-air ducts are of tile and galvanized iron and lead the
cold air from the outside of the building to the stack boxes,
where it is warmed in its passage over the surface of the indirect radiators.

Mixing-Boxes. Mixing boxes will be provided outside of the stack boxes,
for the purpose of regulating the temperature of the air admitted to the room. These are supplied with dampers, conducting the current of air either over the surface of the indirect radiators or up through them, thus regulating the heat.

Registers. Registers will be provided through which the heated air
is admitted to the room over the indirect stacks.

Ventilation. Ventilation is at all times secured by the admission of
outside air through the indirect stacks, thus tempering the
cold air to such a degree as is agreeable.

Pipe in Vent-Flues. Pipe in these flues is provided to increase the circulation
of air. Steam is admitted to these pipes through globe valve.
The heating of these flues by these pipes accelerates the
passage of air, materially increasing their efficiency.

Tile Smoke-Pipe. Tile smoke-pipe is provided for chimney, because this interior surface is round and perfectly smooth, allowing no ac-
cumulations of soot or ashes to impede the draft under any
conditions, and allowing the smoke column to pass rapidly
out in its naturally spiral form, thus giving the best possible
draft obtainable.

STEAM CIRCULATION.*

BY HARELL STEAM HEATING COMPANY.

SPECIFICATION.

Halls and Vestibule—7 feet 3 inches by 38 feet by 10 feet First Floor. 6 inches—containing 2898 cubic feet of space, require to be heated 1 to 35, and will want 83 feet of heating surface. Will place two 40-foot indirect radiators in this hall.

Parlor—15 feet 3 inches by 19 feet by 10 feet 6 inches— contains 3045 cubic feet of space, requires to be heated 1 to 30, and will want 111 feet of heating surface. Will place two 60-foot indirect radiators in this room.

Library—15 feet by 18 inches by 10 feet 6 inches—contains 2835 cubic feet of space, requires to be heated 1 to 30, and will want 94 feet of heating surface. Will place two 50-foot indirect radiators in this room.

Dining-Room—15 feet by 18 feet by 10 feet 6 inches—contains 2919 cubic feet of space, requires to be heated 1 to 30, and will want 97½ feet of heating surface. Will place two 50-foot indirect radiators in this room.

Toilet-Room.—Will place one 6-foot direct radiator in this room.

The location of the registers for all the indirect stacks to all rooms to which they are applied on this floor is substantially shown on the accompanying plan, Fig. 2.

Hall—7 feet 3 inches by 47 feet by 9 feet 6 inches—contains Second Floor 3239 cubic feet of space, requires to be heated 1 to 50, and will want 66 feet of heating surface. Will place in this hall two 34-foot direct radiators.

Chamber No. 1—15 feet by 14 feet by 9 feet 6 inches—contains 1995 cubic feet of space, requires to be heated 1 to 40, and will want 50 feet of heating surface. Will place in this room one 50-foot direct radiator.

Chamber No. 2—15 feet 3 inches by 16 feet by 9 feet 6 inches —contains 2318 cubic feet of space, requires to be heated 1 to 40, and will want 58 feet of heating surface. Will place in this room one 58-foot direct radiator.

Chamber No. 3—15 feet by 18 feet 6 inches by 9 feet 6 inches— contains 2160 cubic feet of space, requires to be heated 1 to 40, and will want 54 feet of heating surface. Will place two 28-foot direct radiators in this room.

* From *The Metal Worker*, May 25, 1889. Copyrighted, 1889, by David Williams.

Chamber No. 4—14 feet by 14 feet by 9 feet 6 inches—contains 1862 cubic feet of space, requires to be heated 1 to 40, and will want 45 feet of heating surface. Will place in this room one 45-foot direct radiator.

Sewing-Room—7 feet by 8 feet by 9 feet 6 inches—contains 532 cubic feet of space, requires to be heated 1 to 40, and will want 14 feet of heating surface. Will place in this room one 14-foot direct radiator.

Bath-Room—5 feet by 9 feet by 9 feet 6 inches—contains 427 cubic feet of space, requires to be heated 1 to 30, and will want 14 feet of heating surface. Will place in this room one 14-foot direct radiator.

The location of radiators and the manner in which they will be connected to risers supplying same on this floor are substantially shown in the accompanying plan, Fig. 3.

Third Floor —Attic. *Billiard-Room*—10 feet by 15 feet by 9 feet, with recess 3 feet by 6 feet by 9 feet—contains 1512 cubic feet of space, requires to be heated 1 to 40, and will want 38 feet of heating surface. Will place in this room one 38-foot direct radiator.

Children's Play-Room—14 feet by 14 feet by 9 feet, with bay recess—contains 2088 cubic feet of space, requires to be heated 1 to 40, and will want 52 feet of heating surface. Will place in this room one 52-foot direct radiator.

Chamber—10 feet by 15 feet by 9 feet—contains 1350 cubic feet of space, requires to be heated 1 to 40, and will want 34 feet of heating surface. Will place in this room one 34-foot direct radiator.

The location of radiators and the manner in which they will be connected to risers supplying them on this floor are also substantially shown in the accompanying plan, Fig. 4.

Boiler. Boiler to be used in this work to be a vertical tubular boiler, made of 60,000 test steel plate ; shell to be $\frac{1}{4}$ inch thick, heads $\frac{3}{8}$ inch thick. The diameter of same will be 45 inches and the hight will be 40 inches, and it will contain 130 2-inch tubes 35 inches long, and 34 3-inch tubes 35 inches long. Total heating surface in boiler, 273 square feet. Actual heating surface in boiler at water-line, 204 square feet ; has an estimated horse-power of 22¾, and actual horse-power of 17. Area of grate surface, 4 square feet. The boiler will be set in brick-work having a cast-iron front of the most improved pattern, containing feed door, clinker door, ash-pit door and automatic draft door. Cast-iron frames with doors are placed in brick-work in their proper places to allow the accumulation of dust and soot to be removed. The boiler will be fitted with one steam-gauge, one water-gauge, three gauge-cocks, one safety-valve, one blow-off cock, one flue-

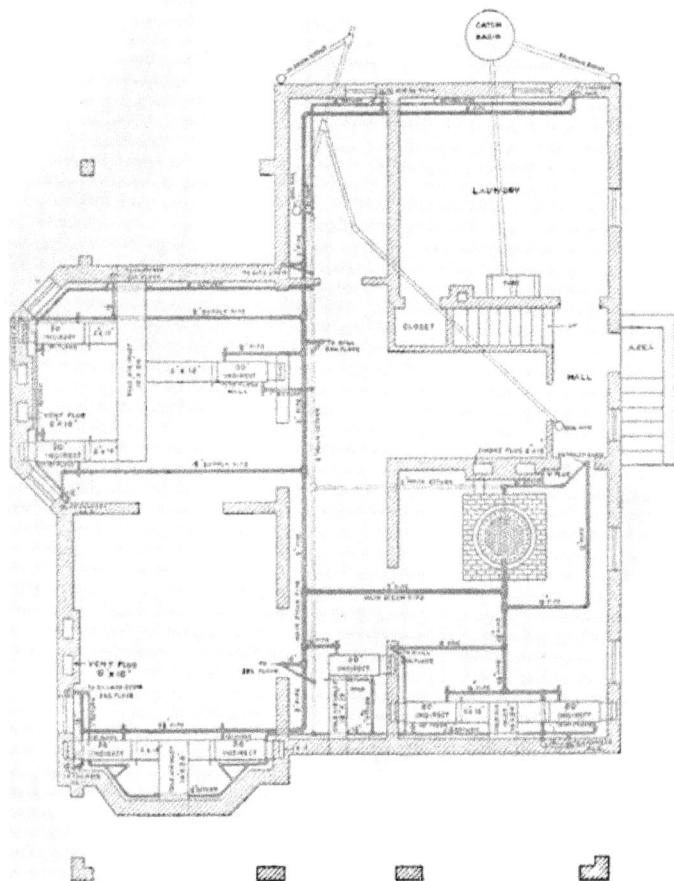

Fig. 1.—Cellar Plan.—Scale, 1-12 Inch to the Foot.

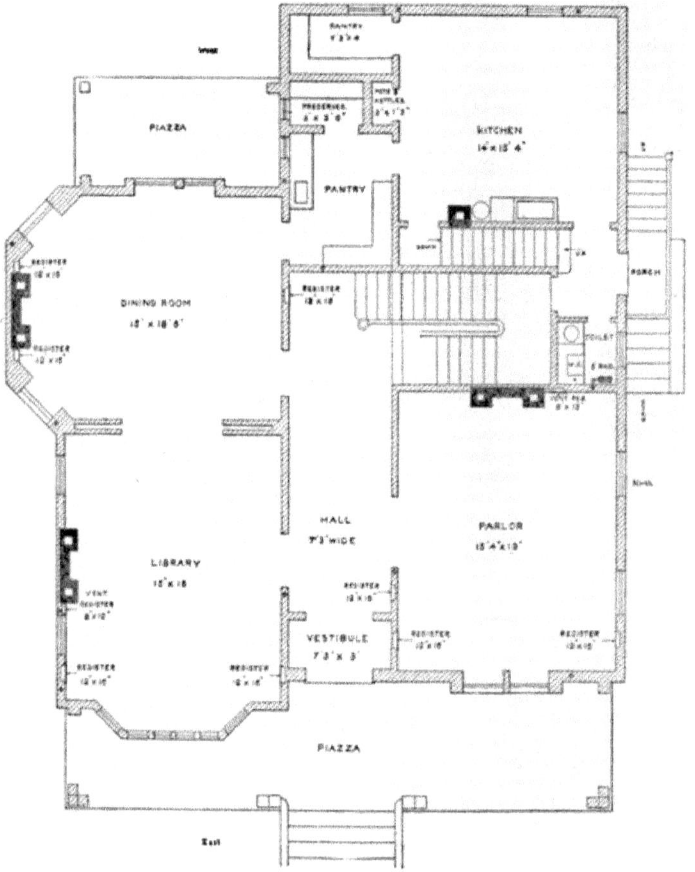

Fig. 2.—First-Floor Plan.—Scale, 1-12 Inch to the Foot.

cleaner, one automatic draft regulator and one automatic water-feeder.

The indirect radiators will be of cast iron and of the best **Radiators.** standard make; the same will be placed under the floors of the various rooms for which they are intended, and will be securely fastened to the joists with wrought-iron hangers; the recess in joist to receive the indirect stacks will be lined with one-cross bright tin. The casings for indirect stacks will be made of No. 24 galvanized iron; the seams in same are to be riveted and soldered. The cold-air ducts for conducting pure air from outside of building to indirect stacks will also be made of No. 24 galvanized iron; with all seams in same riveted and soldered. The openings of cold-air ducts extending through cellar wall to outside of building will be protected with brass wire netting made from No. 9 wire, ¼ inch mesh. There will be one main shut-off valve in each cold-air duct placed in same just inside of cellar wall, and also one valve for each stack placed close to same for the purpose of controlling the admission of air in the stacks, all of which are substantially shown on accompanying plan, Fig. 1.

The conducting or hot-air tubes leading from the indirect **Hot-Air Pipes.** stacks under floors to the registers above will be made from two-cross bright tin, and the sizes of same will be 3¾ x 15 inches, seams to be grooved and soldered, and all of which shall be placed and located as is substantially shown on accompanying plan, Fig. 2.

The direct radiators will be the cast-iron loop radiators of **Direct Radiators.** the best standard make, and will be placed in the several rooms as located on the accompanying plans, Figs. 3 and 4.

All registers used in this work are to be nickel-plated; the **Registers.** hot air registers for indirects will be 12 x 16 inches in size, and those for ventilation will be 8 x 12 inches in size, and will be placed as located on accompanying plans, Figs. 2, 3 and 4.

The manner and mode of piping and the sizes of the **Piping** same are shown on plan of same, and all of which will be run and placed as located on the accompanying plan, Fig. 1.

All rooms in which heat is placed will be ventilated, and **Ventilation.** will have nickel-plated registers 8 x 12 inches in size in each ventilating flue; the same will be placed low down or just above base-boards in all rooms except the smoking and dining rooms, which will be placed near ceilings, and all of which will be located as is shown on accompanying plans, Figs. 2, 3, 4 and 5.

All the risers conveying steam for radiators on second **Risers.** and third floors will be not less than 1¼ inches in diameter, and be run up between the studding in the walls and partitions, and will be thoroughly covered and protected by the best non-conducting material known; connections to radiators will be taken from them at the proper hight to conform with the hight of radiator connections from floor, and come in through the base-board to radiator and thus avoid the cutting of carpets if brought through the floor. Radiators will be connected with union couplings having ground joints, and which admit of their being disconnected readily and taken away if so desired at will, all of which are substantially shown on accompanying plan, Fig. 5.

Valves. All valves on indirect stacks to be made of best steam metal, to have vulcanized asbestos seats and wood wheels. All valves on direct radiators to be finished and nickel-plated, to have vulcanized asbestos seats, ground-joint union couplings and wood wheels. Air-valves on indirect radiators to be of a style known as the automatic air-valves. On direct radiators to be of a style known as the automatic and nickel-plated.

Pipe and Fittings. All pipe used in this work to be of the best quality, all fittings used to be the best gray iron cast fittings.

Floor Plates. Where each supply-pipe to radiator comes through base-board, the same shall be provided with a nickel-plated floor plate.

Pipe Covering. All risers and main and return pipes in cellar will be covered with mineral wool steam-pipe coverings.

Automatic Water-Feeder. The boiler will be equipped with an automatic water-feeder, one that is effective in operation and always reliable.

Automatic Heat-Regulating Apparatus. The entire heating system will be controlled by an electric automatic heat-regulating device whereby the temperature in any apartment can be maintained at a given degree. The system used will be one that is quick in operation and always reliable.

Bronzing. All direct radiators will be bronzed with No. 6000 pale gold bronze.

ADVANTAGES AND GENERAL FEATURES.

You will note that we have deviated somewhat from the cuts furnished us by you, having increased them to ¼ size. We could not show the plant intelligently, or as it ought to be shown in a case of this kind, on a ⅛-inch scale. Our plans are very complete, showing every pipe and how run, also giving sizes and locations of same.

The first floor we heat entirely by indirect radiation for the following reasons : By heating in this way the air is received from outside of buildings into radiator stack; there it becomes heated and passes off through the registers into the various rooms in which they are placed, and if the cold-air ducts and the casings of the radiators are

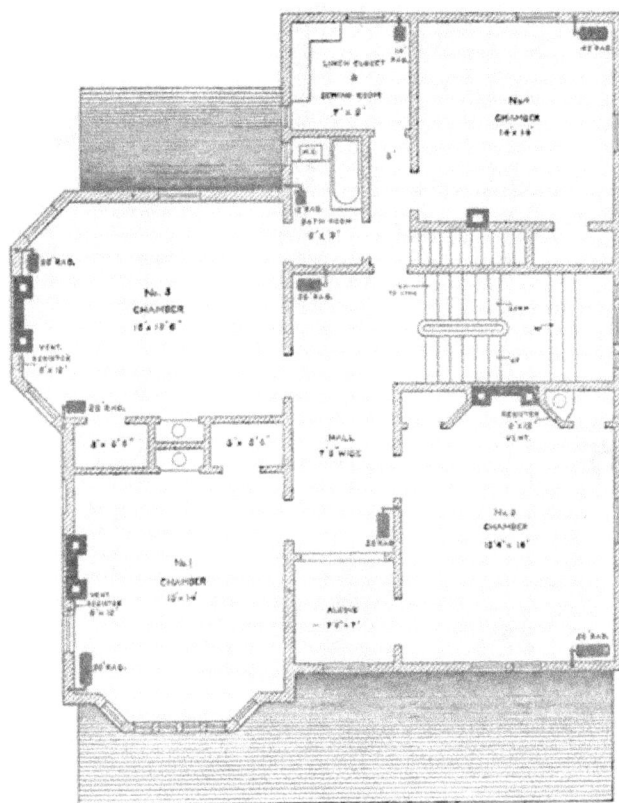

Fig. 3.—Second-Floor Plan.—Scale, 1-12 Inch to the Foot.

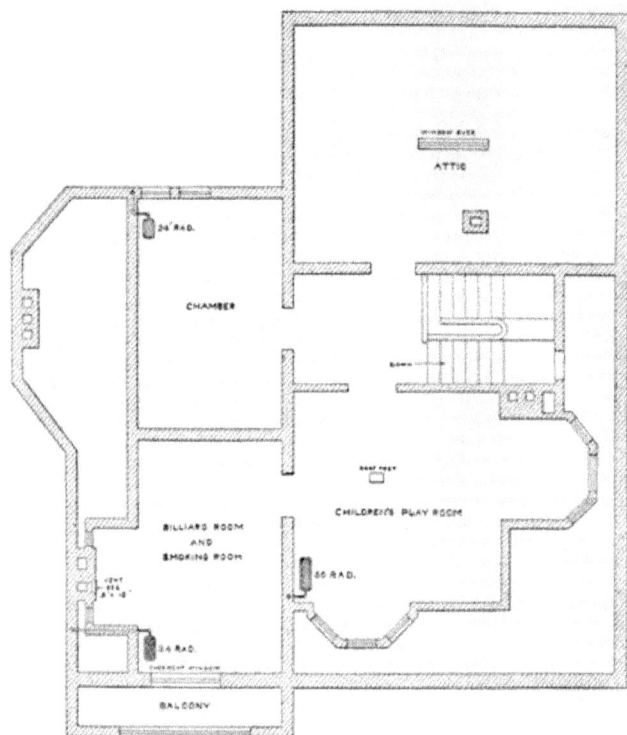

Fig. 4.—Attic Plan.

made tight (and hence we use galvanized iron for this purpose), so that there will be no chance for the foul air in cellar to get up into the stack, we certainly must receive our heated air into those rooms in as pure a condition as it is possible to get it. We also, in addition to this, do not take up any space in any of the rooms in which it is placed and which also can certainly be credited in its favor, and while we admit that it is a more expensive way of heating than the direct system, yet we think the advantages we have named that it possesses over the direct system more than overbalance the difference in cost, and as this is the living part of the house, it requires to be as perfect as possible. This floor is fitted up at the ratio of 1 to 30, or 1 foot of heating surface to 30 feet of space, and which we guarantee will keep the temperature at 70° in all conditions of weather with a steam pressure not exceeding 3 pounds at boiler.

The second and third floors we have fitted with direct radiators, and in doing this we felt that it would fully meet the requirements made upon it, as heat will not be required from the radiators located in the rooms much more than half the time that the apparatus is run during the season, and even then when it is used generally through the day and not during the night when the rooms were occupied. Those rooms receive the benefit of the heat from both the lower and upper halls, and with the thermometer at zero only about one-half of the radiation in the rooms would be required to keep them at the required temperature—namely, 65°—and only when the thermometer ranged from 20° to 25° below zero would the full radiation be required. The first floor being fitted throughout with indirect, the heat of course circulates through the whole house, and every part of the house is receiving more or less of the pure heated air and especially so from the hall radiators, and which we claim is sufficient ventilation for a house of this size, and this being the case, we claim that direct radiation on the second and third floors is preferable to indirect, from the fact that it is less expensive to put in, and that it is more cheaply run, as it is a well-known fact that indirect radiation consumes about one-third more fuel than direct, and while the system is perfect and complete with the direct radiation on these two floors, that indirect could not make it more so. The second and third floors we have fitted up at the ratio of 1 to 40—1 foot of heating surface to 40 feet of space. The plant will be furnished with an electric heat-regulating apparatus whereby the temperature of any room in the house where heat is applied can be controlled and kept to any temperature desired. This apparatus will be one that will be guaranteed to work perfectly and to give entire satisfaction.

The boiler employed in this work and called for in the specification is of ample size, and will do this work very economically. It has an actual heating surface of 204 square feet that cannot be disputed.

Every inch of this is either at or below the water-line. Our formula for heating boilers of this description is 1 foot of heating surface in boiler to 6 feet of radiating surface. There are 835 feet of radiation in the building outside of the mains, and with the mains properly covered with a good heat non-conducting covering the loss from same would not be equal to more than 50 feet of radiation. This makes the total amount of radiation in building 885 feet. The capacity of boiler at the low rate is 1224; this leaves a margin of capacity in favor of the boiler of over 25 per cent. The boiler is furnished with an automatic draft regulator and automatic water-feeder. For all boilers that are designed to be used in this class of work and especially for dwellings it is absolutely necessary that they should be as nearly automatic in all their workings as it is possible to make them with safety and economy; for instance, a reliable automatic water-feeder is very desirable on boilers of this description, for in many cases they are apt to be neglected, and the automatic feeder proves itself very essential at such times, and especially again in case a water-gauge glass should break the automatic feeder will take care of the boiler and prevent its being burned, as it will continue to supply the boiler until the accident is discovered; and again it keeps the boiler regularly supplied up to the proper working hight, thus preventing the possibility of there being either too much or too little water in the boiler at any time, for the nearer the water in the boiler is kept up to its proper working level the better the boiler will work. The ventilation in this building is also very properly arranged, being ventilated from the lowest points in all the rooms save the dining and smoking rooms. Those we have ventilated from above, as, for instance, we wish to free the smoking-room from the smoke as rapidly as possible, and not bring it down and around the persons in the room on its way out; this applies also to the dining-room. All other rooms in the dwelling we claim ought to be ventilated from below, as all foul or vitiated air, being heavier than the pure air, descends to the floor; therefore, the means for its escape want to be at this point, and the arrangements and locations of such are shown in the accompanying plans, Figs. 2, 3, 4 and 5.

The direct radiators used in this work will be what are known as the cast-iron loop radiators, and the indirects will be cast-iron sectional radiators. We have used in our work all of the different makes of radiators that are manufactured, and our experience has been that the cast-iron radiators have proved more efficient, and have given better satisfaction than the wrought-pipe radiators. We feel that these results are largely due to the metal of which they are composed, and also to the design of same, and which we account for as follows: The cast-iron being of a more porous and open nature than wrought-iron, it radiates heat more freely, and again it will be found that in all first-class standard makes of those radiators the openings or steam-ways in same are much larger than in the wrought-tube radia-

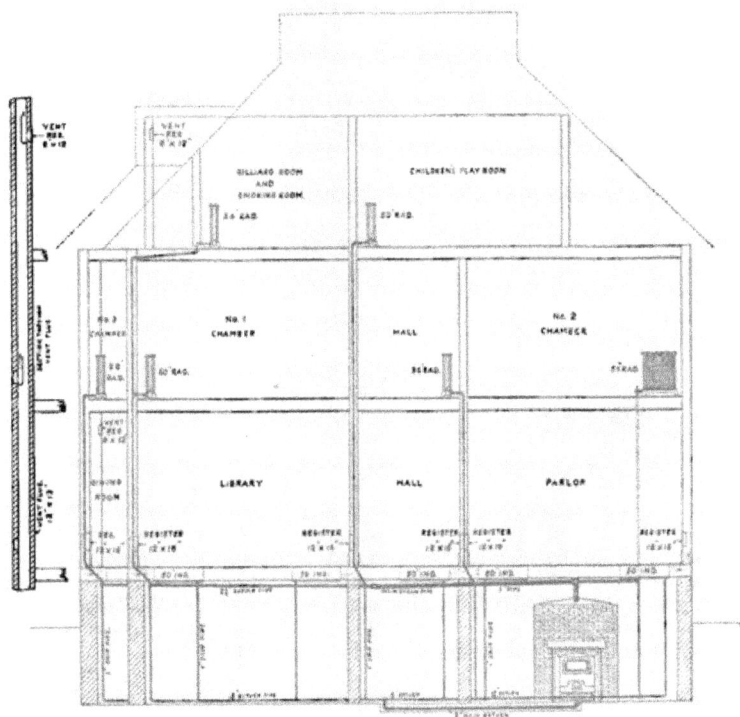

Fig. 5.—Sectional Elevation.—Scale, 1·12 Inch to the Foot.

tors, and which causes the steam to circulate in radiators more freely, and also that the forms of the same are such that it causes it to come more freely in contact with the surface of the same.

The piping employed in this work, and which is shown in accompanying plan of same, is of our regular practice, and which is to drip all mains from the boiler. Drip-pipes are located and connect each riser to main return-pipe in cellar. The second and third floors are fitted with the one-pipe system, which requires the rising pipes and connections to radiators being of sufficient size and so graded as to allow the steam that is condensed in radiator to gravitate back through the pipe to the relief-pipe in cellar, and back through the return to boiler. The reasons why the one-pipe system is preferable to the two-pipe system are that when the work is properly run and graded as spoken of above it is perfectly noiseless in its workings, and also that you receive the condensation back into the boiler again at a higher temperature than you do in the two-pipe system on account of the condense returning back to the water line in cellar through the supply-pipe at nearly the same temperature as the steam, and which causes a more uniform working of the apparatus, as the water of this higher temperature is more quickly converted by the boiler into steam again, and thereby steadier boiler pressures are maintained, and which also makes the apparatus a great deal more economical in fuel; and one more advantage that it has over the two-pipe system is that in shutting off or turning on a radiator you have only one valve to manipulate; therefore it matters not at what pressure you are working under, you have but one valve to open or close, as the case may be, while in the other case you have two, and if working under any pressure, say 3 to 5 pounds, in shutting off a radiator unless you close the return-valve first, or before you close the steam-valve, the water will be forced up into the radiator from the boiler, which not only makes an unbearable noise, but also makes it dangerous in having this water leave the boiler, and especially is this the case where automatic boiler-feeders are not used. The indirect radiators will be fitted with return-pipes, as it is not practicable to run those successfully with the one-pipe system.

The steam or supply valves for the direct radiators will be made of the best steam metal and of the latest and most improved pattern, and are to be thoroughly nickel-plated, and are to be fitted with wood wheels and ground joint union couplings to connect to radiators; they will also be fitted with adjustable seats which can be removed in case of breakage or other causes, and new seats placed in without removing the valves from their places; seats to be of patent metal, and of which there are several makers. The advantage of those valves over the ordinary valves is that the seat of the valves of this description is about the only part that ever gives out. As the main or body part will last almost indefinitely, and by the placing in of a new seat the valve becomes again comparatively new, and while the seat is composed of material that is softer than the valve which sets into it, yet the nature of it is such that it has good wearing qualities and does not destroy the face of the valve, and therefore is in reality the only part of the valve which can give out, and which can be replaced at a very trifling cost. The valves for the indirect radiators will be the same as the directs, only that they will not be nickel-plated.

The air or vent valves for radiators will be of the automatic kind, requiring no attention from the occupants of the dwelling, and of which there are several makers, and by the use of which the radiators are always working. The preference for the automatic over the adjustable valves is that with the adjustable, in case of the steam pressure going down sufficient to cause a cooling off of the radiators, then they become filled with cold air, and which, even upon the return of steam or pressure the attendant has to open the valve to allow the cold air to escape before the radiator will become heated up again, while with the automatic, when once adjusted, it requies no further attention, for its arrangements are such that when the pressure is down and the radiator cooled off the valve opens and remains open until the radiator again becomes filled with steam and heated up, when the valve closes, and the apparatus, having all other automatic devices, would not be complete without this one also.

It is very essential that all exposed pipes, such as the main steam-supply pipes, return and drip pipes, and also the risers to radiators should be covered with a good non-conducting covering, for the reasons that the steam is conveyed to all the points of radiation at a higher temperature, and is therefore more efficient at such points than when pipes are exposed to the atmosphere, and although while all the radiation and condensation in the pipes are not prevented, yet they are reduced to a minimum, and which also makes a great reduction in the consumption of fuel.

ESTIMATE.

Estimate of cost for steam-heating apparatus as per plans and specifications furnished:

45 feet 3-inch pipe	@ $0.58	$26.10	
21 feet 2½-inch pipe	.44	9.24	
82½ feet 2-inch pipe	.28	23.10	
82½ feet 1½-inch pipe	.22	19.46	
		77.90	
62½ per cent		48.69	
			$29.21
243¼ feet 1¼-inch pipe	.12½	30.43	
179 feet 1-inch pipe	.09½	17.00	
		47.43	
52½ per cent		24.90	
			22.53
1 3-inch gate-valve		16.00	
1 2-inch gate-valve		6.25	
		22.25	
35 per cent		7.79	
			14.46
8 1¼-inch angle-valves	4.00	32.00	
8 1-inch angle-valves	2.80	22.40	
		54.40	
60 per cent		32.64	
			21.76
9 1¼-inch N. P. radiator-valves	5.65	50.85	
3 1-inch N. P. radiator-valves	4.10	12.30	
		63.15	
60 per cent		37.89	
			25.26
Boiler and trimmings	•	400.00	
25 per cent		100.00	
			300.00
Mason work			30.00
32 pipe-hangers	.30	9.60	
1 3-inch elbow		1.10	
2 3-inch T's	1.50	3.00	
1 3 x 2½-inch ⌐		1.75	
1 3 x 2 x 2 T		1.75	
3 3 x 1½ T's	1.75	5.25	
2 3 x 1¼ T's	1.75	3.50	
1 3 x 2½ x 1½ T		1 75	
1 2½ x 1½ T		1.25	
2 2½ x 1¼ T's	1.25	2 50	
3 2 x 1¼ T's	.70	2.10	
Carried forward		$33.55	$443.22

Brought forward....................		$33.55	$443.22
1 2 x 1½ x 1¼ T......................		.70	
5 1¼ x 1 x 1½ T's......................	$.44	2.20	
2 1½ x 1 x 1½ T's......................	.44	.88	
2 1¼ x 1 x 1¼ T's......................	.35	.70	
4 1½ x 1¼ elbows......................	.29	1.16	
3 1½ elbows......................	.25	.75	
1 2 x ¾ T......................		.70	
1 2-inch T......................		.60	
4 2 x 1 T's......................	.70	2.80	
1 2 x 1½ T......................		.70	
1 1½ x 2 T......................		.70	
2 1½ x 1¼ x 1 T's......................	.44	.88	
8 1¼ x 1 T's......................	.35	2.80	
3 1¼ x 1 x 1 T's......................	.35	1.05	
5 1¼ 45° elbows......................	.26	1.30	
44 1-inch elbows......................	.13	5.72	
2 2-inch L's......................	.40	.80	
20 1¼-inch L's......................	.20	4.00	
		52.39	
70 per cent......................		36.67	
			15.72
45 feet 3-inch covering......................	.35	15.75	
21 feet 2½-inch covering......................	.30	6.30	
82½ feet 2-inch covering......................	.25	20.63	
88½ feet 1½-inch covering......................	.24	21.24	
243½ feet 1¼-inch covering......................	.23	56.01	
179 feet 1-inch covering......................	.21	37.59	
		157.52	
30 per cent......................		47.26	
			110.26
Red lead, oil, waste, &c......................			5.00
Incidentals......................			20.00
40 sections indirect radiators......................	1.60		64.00
435 feet direct radiators......................	.28		121.80
7 8 x 12 vent. registers, N. P......................	2.80		19.60
8 12 x 16 hot-air registers, N. P......................	3.50		28.00
Bronze and liquid......................			5.00
12 floor plates......................	.10		1.20
8 automatic air-valves for indirects......................	.60		4.80
12 automatic air-valves for directs......................	.60		7.20
Automatic water-feeder......................			18.00
Electric automatic heat-controlling apparatus			300.00
24 days' work, two men......................	4.00		96.00
Brass wire screens for cold-air ducts......................			8.00
Sheet-tin and galvanized iron for indirects and cold-air ducts, including labor			97.23
Hot-air flues for indirects......................			23.56
			$1,388.59
Profit, 20 per cent......................			277.71
Total......................			$1,666.30

STEAM CIRCULATION.*

BY W. B. GUIMARIN.

SPECIFICATION.

Excavate the room, Fig. 1, under parlor 2 feet by 6 inches Boiler-Room. for boilers. Deaden the noise made in the boiler-room by filling in between the ceiling and floor with mineral wool. Use wire laths for plastering overhead. I select this room for the boiler because I think it is the best to get a flue and for light, and about as convenient as any of the other rooms for coal and water.

The boiler to be a base-burning magazine feeder, with Boiler. drop flues. The boiler shell to be ⅜ inch thick and made of the best flange iron.

Dimensions.—Diameter of shell, 3 feet 6 inches; hight of shell, 4 feet; diameter of fire-pot, 2 feet 6 inches; total hight over all, 5 feet 7 inches; 26 tubes; diameter of tubes, 3 inches; tubes 2 feet 6 inches long; 140 feet heating-surface; 6 square feet area of the grate-surface. Size of chimney-flue, 12 x 14 inches. Set the boiler in brick with an air-space all around Boiler it. The square of the brick-work over all, 6 feet. The hight Setting. of brick-work over all, 6 feet. The grate to be a rocker and to dump easily. Ash-pit 13 inches high.

The Manner of Setting the Boiler and Putting on the Trimmings. —Set the boiler on the ash-pit. Build a 12-inch brick wall, with an air-space of 3 inches between the boiler and brick. Leave hand-holes near bottom for cleaning out the soot. Build up square to the top of the boiler with the flue entering directly into the chimney-flue. Arch over, and in the center fit closely around where the magazine casing comes. The magazine to be cast-iron, and left loose, so that it can be taken out. The fire-door to have a good slant toward the fire-box. The ash-pit door to fit closely, and be regulated by Boiler an automatic diaphragm regulator. This pipe to connect Fittings. below the water-line. The safety-valve pipe to be 2 inches, and connect in the top of the boiler and carried to the nearest point outside. The safety-valve to be brass, with a lever and weight that can be set to suit any pressure. The end of this lever to be attached to a diaphragm, the same kind as the one to regulate the ash-pit door. The pipe from this to extend

---

* From *The Metal Worker*, August 31, 1889. Copyrighted, 1889, by David Williams.

down inside the boiler 8 inches below the water-line. The blow-off to be 1½-inch pipe connected to the bottom of the boiler, with a common plug-cock on the side of the boiler with a hose end. Within 2 or 3 feet of this commence and run a 1½-inch pipe outside, and independent of any other pipe. When you want to blow out the boiler connect this pipe to the cock by means of a hose. When through be sure and disconnect them, because if the blow-off cock should leak you can detect it. Do not under any circumstances connect the blow-off pipe to the sewer. Just back of the blow-off cock between the boiler and cock leave a tee to connect the water-supply from the tank in the attic. The compression water-gauge cocks to be 3 or 4 inches apart, and the lower one to be 3 inches above the top of the flues. The glass water-gauge to have the bottom pipe connected below the water-line. The top pipe near top of the boiler.

Water-Supply. Connect a ¾-inch galvanized-iron pipe to tank in the attic with a straightway-valve close to the tank, run down through the kitchen exposed and in the corner by the range-flue to the laundry, thence across under the ceiling to the boiler, with a straighway-valve in easy reach to feed the boiler with water. Connect this pipe below the water-line or in the blow-off pipe between the blow-off cock and the boiler. This pipe should be independent of any other water-pipe, for the reason that if the water has to be shut off for causes in the plumbing the boiler will not be interfered with. I am opposed to automatic water-feeders, for one less than perfect cannot be relied on. It requires but little more trouble or skill to fill the boiler with water when it is filled with coal. Use anthracite coal.

Trimmings. One 2-inch safety-valve with lever and weight with diaphragm attachment ; one automatic damper-regulator ; one glass water-gauge ; one steam-gauge ; two wooden handles ; compression water-gauge cocks; hight of water-line in boiler, 4 feet 6 inches.

Main Steam-Pipe. The main steam-pipe to be 3½ inches, run directly up from boiler within 12 inches of the ceiling. Put a 3½-inch brass angle-valve where the turn is made in place of the elbow, thence to south side of the house without reducing.

Flow-Pipes. In the hallway have a cross with the outlets up and down. The upper outlet to be 2½ inches. In this put a short nipple with a 2½-inch tee, the ends pointing with the hallway. Run to the east and west ends of the hall a 2½-inch pipe, the lower outlet to be ¾ inch. In this put a ¾-inch pipe and run down and connect it in the main return. This will get rid of the condensed water at this point. In the

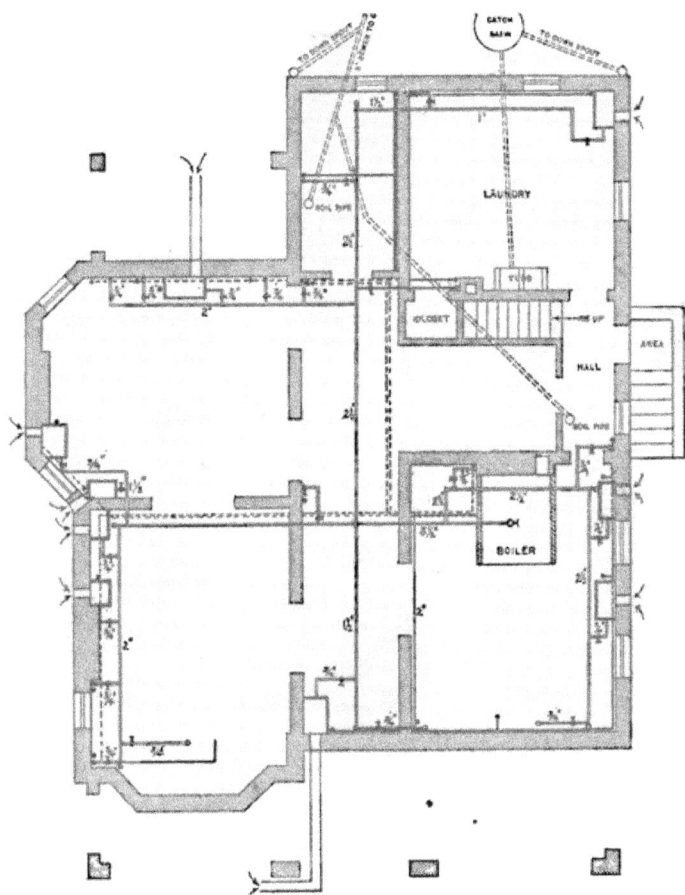

Fig. 1.—Cellar Plan.—Scale, 1-12 Inch to the Foot.

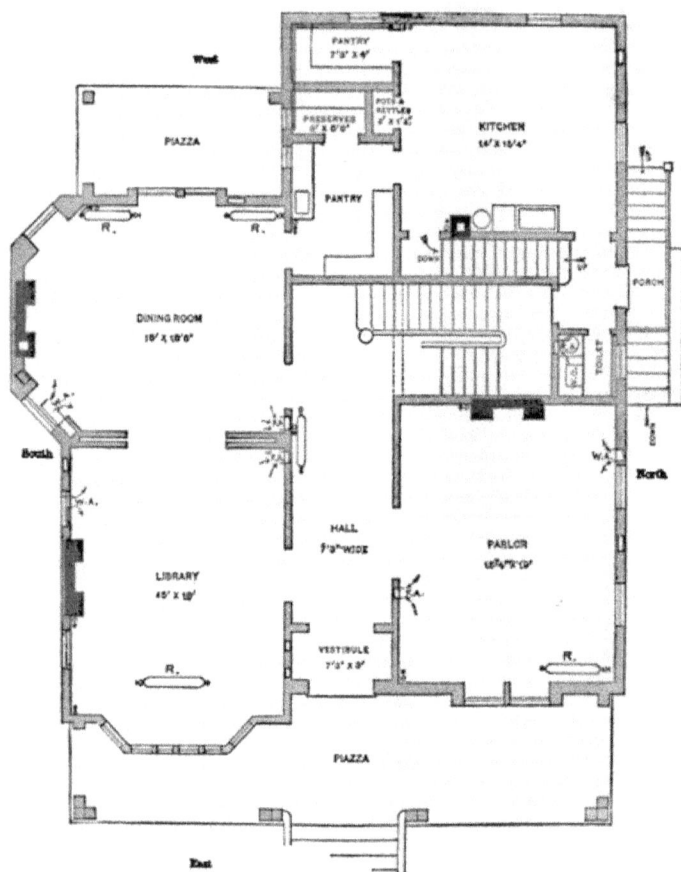

Fig. 2.—First-Floor Plan.—Scale, 1-12 Inch to the Foot.

boiler-room, about 2 feet beyond the angle-valve on the horizontal steam-main, leave a tee with outlet 2 inches, this outlet pointing from side of main steam-pipe. Put in this outlet 2-inch pipe; return back and run this size pipe to and along north side of house to the east side. All the main steam-pipes to incline away from the boiler 1 inch in every 10 feet.

The main return to be run below the cellar floor and be one size smaller and run to correspond with the main steam-pipe. Run back and connect to bottom of the boiler, their extreme ends (away from the boiler) being connected together by the same size pipe as the return. *Return-Pipes.*

The branches that lead off from the main steam-pipe to supply the branches for more than two radiators to be not less than 1½-inch, and this size is not to be for more than four radiators, and all such branches should nipple up and incline from the main steam-pipe, and extend down and connect to the main return. Every radiator to have an independent steam-supply of ¾-inch pipe, except the coil in the bath-room and radiator in the chamber above the bath-room. One supply will do for both of these rooms; also the radiators in the upper and lower hallways—one supply will do for both of these radiators, but every radiator must have an independent return-pipe run down and connected to the return running horizontally on or below cellar floor. The ¾-inch steam-pipe for each radiator must have one ¾ globe-valve for each placed near where they leave the large steam-pipe. The perpendicular steam-pipes where they turn up for the radiator to have a tee fitting looking down into this. Connect a pipe for the relief of the steam riser, then run down and within 12 inches of the cellar floor connect it into its corresponding return riser. All return risers to have straightway-valves placed below this connection and 6 inches above the cellar floor. All return risers and the reliefs for the steam risers to be ¾-inch pipe. The steam-riser valves and the return-riser valves are for the purpose of cutting off entirely these sections of the pipes independent of any others (the steam-valve to cut off the steam, the return-riser valve to prevent water from backing up in the riser). *Branch Flows and Returns.*

The main return-pipes, where they run below the cellar floor, to be out of the way of the crossings and doorways, as shown on cellar plans by dotted lines, to be laid in a cast-iron trough. This trough is to be 8 inches by 8 inches square with a flange on each side, with holes on the flange and made in sections that can be easily handled and to suit the corners and cross sections. The cover to be cast-iron, with holes in *Return-Pipe Troughs.*

it to correspond with the holes in the flange of the trough, so
that it can be screwed down tight. Use brass screws. The
cover is to be in sections that would measure from the center
of one pipe to the center of the next pipe. The holes where
the pipes come through would be a half-circle. The half-
circle should extend up 3 inches and be in the clear 1 inch
larger than the pipe which comes through the cover. The
holes in the flange at the trough should have threads cut in
them, and by placing a gasket of asbestos packing the same
size as the cover, it can be bolted down perfectly water-tight.
The ends of the sections of the trough and cover should be
joined together with red lead made thick into putty. The
top of the cover to be on a level with the cement floor. A
trough and cover made in this way will prevent any water
from getting on the pipes and from washing out the cellar, as
water will have to be 3 inches deep before it will get in the
trough. The pipes to lay on legs of cast-iron placed in the
bottom of the trough. This pipe to have a fall toward the
boiler and have a straightway-valve close to where it connects
to the boiler ; also a swing check in this pipe outside of the
valve. The return-pipes, where they do not come in the way
of the doorways and crossings, to run along the wall 4 inches
above the cellar floor, then connect to the pipe in the cast-
iron box.

Pipe-
Hangers.　　All the pipes near the ceiling to be hung with adjustable
pipe-hangings screwed into the floor joist. Expose all pipes.
Have none in the walls where they cannot be gotten to, bear-
ing in mind that no pipes should be made fast and rigid, but
left loose so they can have easy play for expansion.

Thimbles.　　The pipes, when they go through the floor, to run through
a brass thimble ¼ inch larger than the pipe, with plated
flanges. One end to be threaded, so that the flange can be
screwed on to draw them tight to the ceiling and floor.

All pipes to be wrought-iron of the standard sizes, smooth
and straight and even in thickness.

Pipe
Fittings.　　All pipe-fittings to be heavy cast-iron, smooth and even.

Radiators　　The following radiators to be wrought-iron tubes, with
common return bends and placed as shown on the plans:
Two in the dining-room, Fig. 2. The one nearest the pantry
door to have 30 feet heating-surface, and have a warming-
closet of some neat design attached. The other one 25 feet
heating-surface. The library, 60 feet heating-surface; parlor, 50 feet heating-surface; the lower hall, 30 feet heating-
surface; chamber, Fig. 3, above dining-room, 38 feet heating-
surface; chamber above library, 42 feet heating-surface;
chamber above parlor, 40 feet heating-surface: chamber

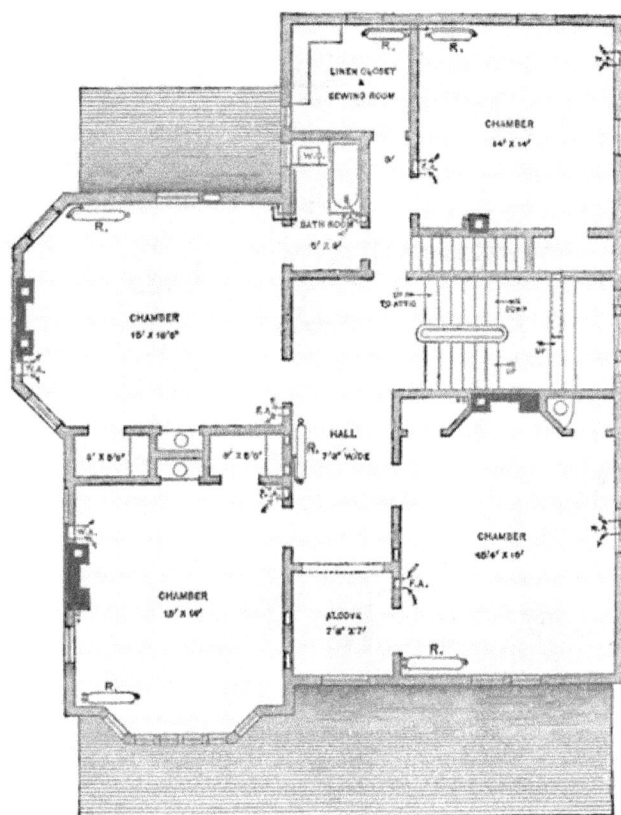

Fig. 3.—Second-Floor Plan.—Scale, 1-12 Inch to the Foot.

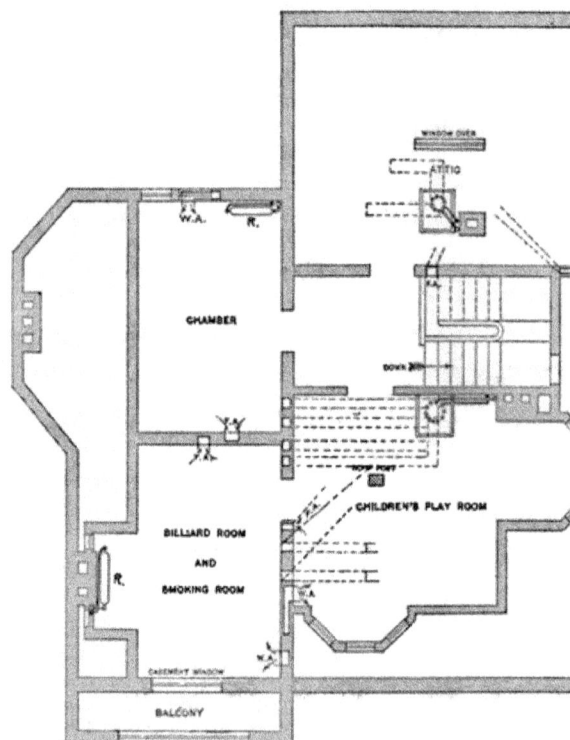

Fig. 4.—Attic Plan.—Scale, 1-12 Inch to the Foot.

above kitchen, 20 feet heating-surface; sewing-room, 10 feet heating-surface; bath-room, 8 feet heating-surface; children's play-room, Fig. 4, 24 feet heating-surface; upper hall, 20 feet heating-surface; billiard-room, 18 feet heating-surface; chamber over bath-room, 12 feet heating-surface; the attic, 20 feet heating-surface.

In the rear of pantry and preserve closet run around the walls, 10 inches above the floor, four 1-inch pipes reaching from window to window, with manifolds placed under each window. Connect the steam-pipe in the upper end of one manifold and the return-pipe in the lower end of the other with valves in each pipe close to the manifolds. Bring the return to the steam-pipe, and run it down the same as a radiator. Connect coil in the bath-room. Run six 1-inch pipes 3½ feet long, with return bends laying flat against the wall, in toilet-room. Run four 1¼-inch pipes underneath window with return bends the same way as in bath-room. *Wall Coils.*

The radiator in the attic, Fig. 4, to be circular, and fit closely around the ventilating-flue, and to assist in exhausting the air from the rooms by heating the flue. Its main purpose is to keep the water in the tank from freezing. The tank can be inclosed to suit this radiator. The radiator in the children's room to be circular, and fit closely around the ventilating-flue. *Attic Radiators.*

The radiators placed along the cellar wall, Fig. 1, are to warm the incoming air for ventilation and are not to be more than 30 inches below the ceiling, to be cast-iron, and their surface to be of the kind and so arranged that the air will be retarded and broken up. Inclose them with Russia iron by forming around the ends and side two sheets and riveting them together so as to have a space of 1 inch between them and form a flange all around so as to fasten to the ceiling and wall. Fill this space with asbestos tightly packed in. Make the bottom funnel-shaped and run through the wall a 10 x 10 inch square pipe as near to the bottom of the heater as possible, with a damper against the wall that can be slid in and out to regulate the air when necessary. On the outside flare out the pipe to receive an iron grating 12 x 12. This bottom piece to have a flange and bolted tight to side and ends. A box made in this manner will keep the cellar cooler and can be easily taken away from the heater when necessary. *Cellar Radiators for Ventilation.*

Cover the pipes with removable asbestos covering in the rooms to be kept cool for vegetables. *Pipe Covering.*

All radiators to have automatic air-valves of approved make placed on the ends where the return is connected. Run *Air-Valves.*

down with ⅜-inch pipes, all being connected to a ¾-inch pipe, this pipe carried and ending over the nearest sink. I am opposed to connecting the air-valve pipes with the sewer for the reason that when the radiators are not in use there is a chance for the foul air in the sewer to escape through them.

Radiator Fittings.
The cast-iron bases of the radiators to have their legs 6 inches long. This will give full access for sweeping and dusting under them. Each wrought-iron radiator to have two nickel-plated ¾ inch angle-valves with wooden handles, one for steam, the other for the return. Their handles to be straight up.

Pipe Unions.
Unions and right and left sockets to be placed on pipe where they can be easily taken apart when necessary. Pipes 1½ inches and larger to have flange unions. Pipes 1¼ inches and smaller to have right and left sockets. No swivel unions to be used.

Pipe and Radiator Finishing.
All pipes in basement to be painted with black Japan varnish. The radiator and pipes above cellar to be bronzed with bright gold. Tops of the radiators to be cast-iron, of some neat design and bronzed with bright gold.

Position of Radiators.
I place all the direct radiators near the coldest places in the rooms, except the library, Fig. 2, for the reason that all the corners and walls in this room will want to be used.

I do not attempt to ornament or use anything that is fancy, as any radiator can be made ornamental by a metal covering, which design I think the occupant or the lady of the house should choose and pick out.

I have not attempted to use anything automatic to cause complication, for I believe the plainest and the simplest steam job, properly and substantially done, is the best.

VENTILATION.

Air-Pipes.
All the air-pipes and flues to be of good, bright tin, and well soldered, except the round flues which go through the roof. They will be of galvanized iron, well riveted.

Ventilating-Flues.
Each of the following rooms to have a separate cold-air pipe with heater and warm-air flue and foul-air flue: Dining-room, parlor, library and all the chambers. There will be one cold-air pipe and one heater to furnish the two warm-air flues leading to children's play-room and billiard-room, but each of these rooms to have a separate foul-air flue. The following rooms to have foul-air flues and not warm-air flues: Toilet, bath-room and the hall.

The cold or fresh air pipes to be 10 inches by 10 inches square, run directly through the wall, with an iron grating 12 inches by 12 inches on the outside. Connect these as close to the bottom of the heater as possible with a sliding damper against the wall on the inside. This is for regulating or closing off the air.

The fresh-air pipes which pass through the wall under the porches must extend across and take air fresh from the outside, as there is reason to suppose that the air underneath the porch is not so pure as the outer air.

The warm fresh-air flues to connect in the tops of the heaters turned up in the wall between the studding of the outside walls, extend up and opening out into the rooms above the baseboard. *Warm-Air Flues.*

All the warm-air flues of first floor to be 8 inches by 12 inches, with register-valves 12 inches by 14 inches; of the second story 8 inches by 10 inches, with register-valves 10 inches by 12 inches; of the third story 6 inches by 10 inches, with register-valves 10 inches by 10 inches, except the children's play-room and billiard-room. These will be 4 inches by 10 inches, with register-valves 8 inches by 10 inches. All the foul-air flues to start at the baseboard and run up between the studding in the inner walls with a register-valve close to the ceiling and one at baseboard. All of the foul-air flues to be 4 inches by 10 inches, with register-valves 8 inches by 10 inches. The valves close to the ceilings to have an endless cord to hang down and be in easy reach to open and close the valves when necessary. *First-Floor Air-Flues.*

The foul-air flues which lead from the dining-room and library and the chambers above each of these rooms to be carried along under the children's play-room floor between the joists, brought close together and coming through and flanged over the floor close to the chimney in this room. Over these pipes build a square galvanized sheet-iron box large enough to cover and receive them and be fastened tight to the floor in the top and center of this box. Rivet and solder on a galvanized-iron pipe 15 inches diameter. Run this pipe straight through the roof, with a ventilating-cap on the top large enough to prevent any blow-down or raining in Build a wooden platform over the box to protect it from abuse and for the heater to set on. The pipe and box to be made from No. 18 galvanized iron. *Ventilating Flues.*

The foul-air flues from the parlor and chamber above the parlor and the children's play-room to run up above the ceiling of the children's play-room, thence across and connect separately into the 15-inch pipe before it goes through the roof.

The foul-air flues for the billiard-room and the chamber on the same floor adjoining it to be joined together above their ceiling in an 8-inch galvanized sheet-iron pipe, carried straight up and through the roof, with a ventilating cap on it. The foul-air flues for the toilet, the bath-room and chamber above the kitchen to be brought together under the attic-room floor, and made in every particular the same as in the children's play-room, except the size of the galvanized-iron pipe, which is 12 inches in diameter. The ventilation from the hallway to connect in this pipe near where it goes through the roof.

System of Ventilating. I have named three distinct air-flues for the purpose that the system of ventilation may be more easily understood and put in. The cold-air flue is the pipe bringing the outer air to the bottom of the heater. The warm-air flues are the pipes to carry the fresh air in the rooms after being warmed by the heater. The foul-air flues are the pipes leading to the top of the house and through the roof to cause circulation, creating a change of air in the room.

Warm-Air Flues. The warm air should be brought in the room under or near a window of the outside wall, and as these are the places for the coldest air in a room it is nothing but right that it should be met and broken up by the warm air and more natural for the aid of circulation than if they were brought up in an inside wall, hence my reason for bringing them up in the outside wall.

Warm-Air Registers. The warm-air registers are placed above the floor in the wall for the reason that when they are let in the floor they become a good place to catch the sweepings.

Foul-Air Flues. The foul-air flues are carried up in the inside walls for the reason that they will be warmer, which will aid the expelling of the air from the rooms. These registers being at the floor and the ceiling will cause the heavier air along the floor to be displaced, as well as the lighter air at the ceiling.

Indirect Radiators. The amount of square feet heating-surface for each indirect radiator for warming the air for ventilation is as follows : Dining-room, 40 feet heating-surface ; library, 40 feet heating-surface ; parlor, 40 feet heating-surface ; chamber above dining-room, 35 feet heating-surface ; chamber above library, 35 feet heating-surface ; chamber above parlor, 40 feet heating-surface ; chamber above kitchen, 35 feet heating-surface ; chamber in third story, 30 feet heating-surface ; children's play-room and billiard-room, 40 feet heating-surface.

Heating System. The system I employ is the gravity return, which can be run at any pressure, but it will never be necessary to run higher than with 10 pounds. The heating of the house is by direct radiation, and the heating-surface proportioned as though no currents of air were passing through—that is, the rooms are kept to 70° by direct radiators, regardless of ventilation. As the air which enters the rooms is near the same temperature, the changing of the air will not materially affect the temperature of the rooms.

In moderately cold weather the direct radiators can be entirely closed off, as the warm air will keep the house pleasant by closing the foul-air registers where there is an open fire-place.

All registers to be plain nickel-plated.

Estimate.

1 magazine feeder wrought-iron boiler, with trimmings, complete		$380.00
2000 bricks		24.00
36 feet 3½ pipe	@ $0.30	10.00
2 3½ x 2½ T's	1.10	2.20
1 3½ cross 2½ x ¾		1.50
1 3½ to 2½ elbow		.75
1 3½ x 1 inch T		.75
1 3½ brass angle-valve		8.00
3 3½ flange unions	1.25	3.75
3 3½ adjustable hange.s	1.00	3.00
130 feet 2½ pipe	.20	2.60
2 2½ T's	.52	1.00
6 2½ elbows	.30	1.80
6 2½ to 2 inch elbows	.30	1.80
5 2½ x 2 inch T's	.60	3.00
2 2½ to 1½ T's	.40	.80
20 2½ x ¾ T's	.40	8 00
6 2½ flange unions	.70	4.20
1 2½ brass swing-check		6.00
12 2½ adjustable hangers	.40	4.80
63 feet cast-iron trough	1.60	37.80
140 feet 2-inch pipe	.14	19.60
6 2-inch elbows	.20	1.20
6 2-inch to 1½ ells	.20	1.20
20 2 x ¾ T's	.20	4.00
75 1½ pipe	.10	7.50
6 1½ elbows	.10	.60
3 1½ to ¾ elbows	.10	.30
2 1½ to 1¼ elbows	.10	.20
2 1½ x 1¼ T's	.10	.20
2 1½ to 1 inch elbows	.10	.20
18 1½ x ¾ T's	.10	1.80
20 feet 1¼ pipe	.07	1.40
2 1¼ elbows	.08	.16
6 2-inch adjustable hinges	.30	1.80
4 1½-inch " "	.20	.80
6 1-inch " "	.12	.72
30 ¾-inch " "	.10	3.00
35 feet 1-inch pipe	.05	1.75
6 1-inch elbows	.05	.30
Amount carried forward		$552.48

Amount brought forward		$552.48
6 1-inch to ¾ ells	@ $0.05	.30
6 1-inch by ¾ T's	.05	.30
830 feet ¾-inch pipe	.04	32.20
400 feet ¾ elbows	.04	16.00
260 ¾ T's	.07	18.20
200 ¾-inch R and L sockets	.04	8.00
4 1-inch " "	.08	.32
2 1¼-inch " "	.10	.20
8 1½-inch flange unions	.60	4.80
8 2-inch " "	.70	5.60
1 2½-inch brass gate-valve		6.00
31 ¾ globe valves	.70	21.70
26 ¾ straightway valves	.80	20.80
60 feet ¾ galvanized-iron water-pipe	.08	4.80
8 ¾ galvanized-iron elbows	.08	.64
2 ¾ galvanized unions	.30	.60
1 1½ blow-off cock with hose end		2.25
3 feet 1½ hose	.50	1.50
465 square feet heating-surface common return bend radiators	.36	167.40
335 square feet heating-surface cast-iron radiators	.30	100.50
9 sheet-iron boxes for the cast-iron radiators	10.00	90.00
9 iron gratings	.60	5.40
430 feet tin air-flues	.12½	53.75
3 12 x 14 registers	1.50	4.50
4 10 x 12 registers	1.30	5.20
3 10 x 10 registers	1.10	3.30
23 8 x 10 registers	1.00	23.00
90 feet of cord	.10	9.00
16 feet galvanized sheet-iron pipe	1.00	16.00
16 feet galvanized sheet-iron pipe	.75	12.00
6 feet galvanized sheet-iron pipe	.20	1.20
1 15-inch ventilating cap		5.00
1 12-inch ventilating cap		3.00
1 8-inch ventilating cap		1.00
2 galvanized sheet-iron boxes	2.50	5.00
30 wooden handle nickel-plated angle radiator-valves	2.00	60.00
24 automatic air-vents	1.50	36.00
200 feet ⅜ pipe	.02	4.00
60 ⅜ elbows	.02	1.20
30 ¾ x ⅜ T's	.03	.90
10 ¾ to ⅜ ells	.03	.30
40 brass thimbles with nickel-plated flanges	.80	32.00
Gold bronze		5.00
Black varnish		2.00
52 days for one steam-fitter and two helpers	6.00	312.00
18 days for tinner	3.00	54.00
Cartage and incidental expenses		80.00
Total		$1,790.34

STEAM CIRCULATION.*

BY ANSON W. BURCHARD.

SPECIFICATION

The importance of ventilation is universally acknowledged, Ventilation. and the connection of the heating of a house with its ventilation is so inseparable that no heating apparatus which does not combine with it as thorough a system of ventilation as practicable can be considered complete.

Every one who has had occasion to examine the subject knows that very few buildings are provided with efficient means of ventilation, and that however well the heating apparatus may be calculated to maintain the temperature at the desired degree in the coldest weather, in very few cases does it insure in connection with this an abundant supply of fresh air.

Where the question of expense and attendance does not enter the problem, to secure this supply of fresh air is not such a difficult matter, and many large buildings are fitted with appliances for this purpose of a very complete description. Such apparatus require the entire attention of an engineer and are not, therefore, practicable for use in a private residence, such as the one under consideration.

Hence, in designing an apparatus for such a house any ventilating appliances that are adopted should be automatic and the movement of the air induced by the natural drafts of chimneys and hot flues, fans and other mechanical devices being impracticable because of the attention required to keep them in operation. In arranging appliances to afford this ventilation one of the first points to be decided is how much fresh air will be furnished.

Perfect ventilation may be said to have been secured in an inhabited room only when any and every person in that room takes into his lungs at each respiration air of the same composition as that surrounding the building and no part of which has recently been in his own lungs or those of his neighbors or consists of products of combustion generated in the building, while at the same time he feels no currents or

---

* From *The Metal Worker*, October 12, 1889. Copyrighted, 1889, by David Williams.

drafts of air, and is perfectly comfortable as regards temperature, being neither too hot nor too cold.

Very rarely can such ventilation be secured if the number of occupants of a room exceeds two or three.

Air Impurities. Without entering into a discussion of the methods and expense of securing perfect ventilation, good ordinary ventilation is to be secured by keeping the vitiated air diluted to a certain standard. All air with which ventilating appliances have to deal contains more or less impurities, some of which are more dangerous than others and are less affected by this process of dilution. Of these impurities carbonic-acid gas is popularly supposed to be the most harmful, but as a matter of fact it is not poisonous and produces no harmful effect, even when present in 30 to 50 times the normal quantity. But this carbonic acid is generally found accompanied by other gases which are harmful, particularly carbonic oxide and sulphureted hydrogen. Hence, as there is no convenient method of determining the percentage in which these two latter gases are present, it is usual to determine the percentage of carbonic acid present, for which there is a simple method, and assume that the amounts of the other gases present are proportionate to this.

The Odor Test. As a rule an apartment may be considered well ventilated when a person entering it from the fresh outer air does not perceive any special odor, and experience has shown that a faint, musty, unpleasant odor is perceptible under such circumstances if the amount of carbonic acid, of which the normal is about 4 parts in 10,000 of air, be increased to above 7 parts in 10 000. If the air which has been used and contaminated did not mix with the air in the room a comparatively small quantity of fresh air would be required. Basing their estimates on this erroneous assumption, some authorities have concluded that 250 cubic feet of air per hour for each occupant is all that would be required.

Fresh-Air Supply. But as the contaminated air does mix with the fresh air it is found that in order to keep the carbonic acid diluted to 7 parts in 10,000 of air a supply of about 3000 cubic feet of fresh air per hour is needed for each occupant where rooms are occupied continuously.

The rooms of dwelling-houses are rarely occupied continuously, and there is a large amount of air constantly being admitted through the accidental openings, as doors, windows, cracks, &c., so that a supply of 2000 cubic feet per hour for each occupant is sufficient. The house for which this apparatus is designed would have a supply of fresh air as follows, assuming the velocity of air in the flues to be 3 feet per second :

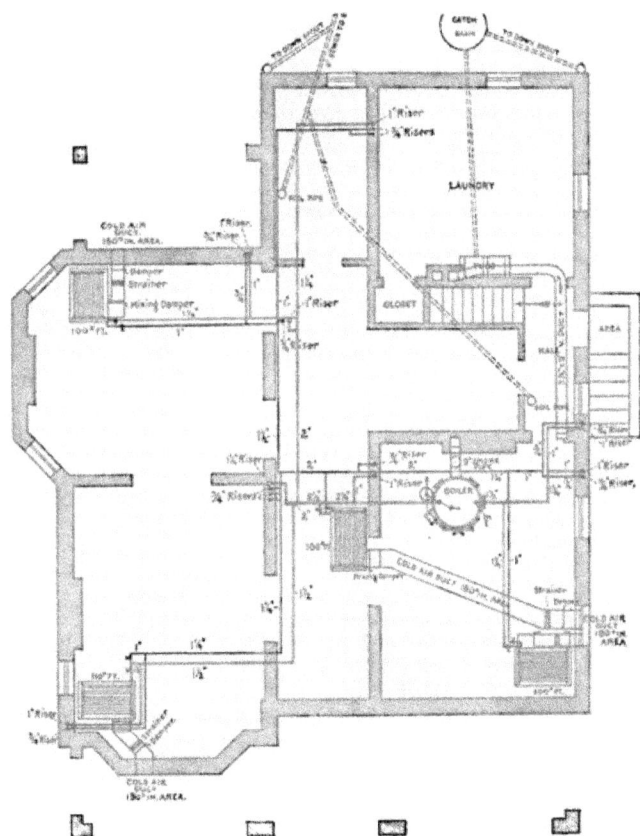

Fig. 1.—Plan of Cellar.—Scale, 1-12 Inch to the Foot.

FLOOR PLANS ACCOMPANYING ESSAY OF ANSON W. BURCHARD.

Velocity of Air Circulation. Experiments with the style of indirect radiators which are proposed in this design have shown a fair average of the velocity of movement of the air through the coils to be 2 feet per second.

Area of Register Openings. The area of the opening of the register in the hall, parlor and dining-room is 205 square inches each, and that in the library 225, giving a supply of 10,500 cubic feet per hour to the three first apartments and a supply of 11,000 cubic feet to the last mentioned, a total of 42,500 cubic feet, enough for over 20 persons. Occasionally a larger number of people than this might be gathered together in the house, but it is not advisable to provide so expensive and cumbersome an apparatus as would be needed to insure an abundant supply of fresh air on such extraordinary occasions. Again, as the rooms on the

Size of Apparatus. second floor would not be occupied continuously by a number of people for a great length of time, it would not be advisable in this case to provide so elaborate and expensive an apparatus as would be required in order to give each room an independent supply of fresh air. The air of those rooms is being drawn out through the fire-places continually, and it is recommended that transoms or other means be provided so that this air may be replaced by fresh air from the halls. The ventilation afforded in this way, although not so thorough as that insured by a system having an independent supply of fresh air for each room, as one employing indirect radiation exclusively, gives very good results practically, and, taking everything into consideration, would prove the most satisfactory.

Temperature As the movement of the air is due to its expansion by the heat, it follows that the higher the degree to which the air is warmed the greater will be its velocity of movement and the greater the quantity which will pass through the registers and enter the room in a given time.

The velocities of movement of the air assumed in the above estimate would not be maintained unless the air were heated to a degree above that required to keep the room warm in mild weather—that is to say, in mild weather the supply of heat, and consequently of air, must be reduced or the rooms become overheated.

Mixing-Dampers. Also when the gas is lighted in a room a considerable amount of heat is given off, for which some allowance must be made in the supply of heat from the radiators. On this account the lack of sufficient ventilation is more frequently noticed in the evening. Therefore if the estimated volume of fresh air is to be introduced into the rooms continuously some provision must be made to lower its temperature in mild weather to prevent this overheating. It is proposed to make

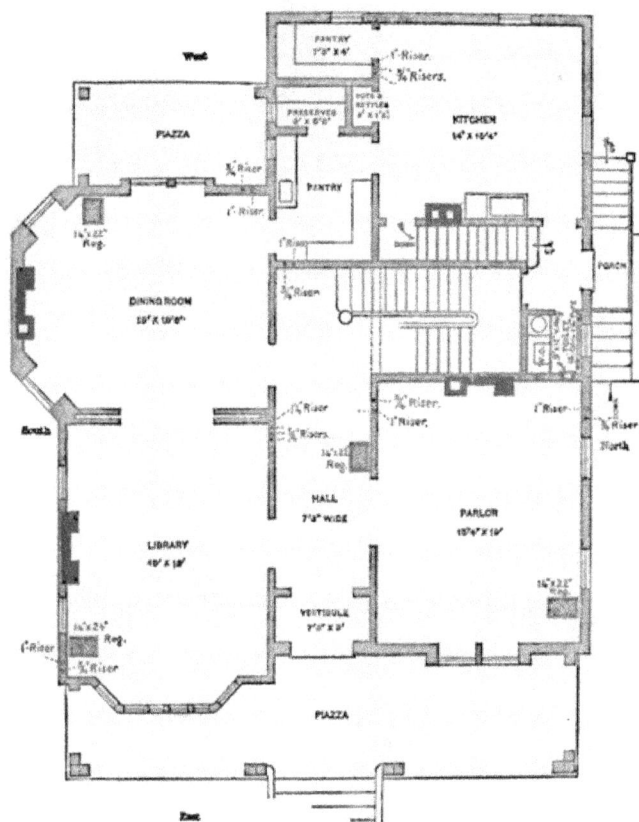

Fig. 2.—*Plan of First Floor.—Scale, 1-12 Inch to the Foot.*

use of appliances known as "mixing-dampers " for this purpose. These are arrangements of ducts and flues such that by the movement of the dampers the air can at pleasure be made to pass either wholly in contact with the radiating-surfaces or wholly separate from them, or partly one way and partly in the other, in such proportions as may be needed to give the air which is discharged into the room the desired temperature.

Where no automatic regulator is used these mixing-dampers are operated by a suitable arrangement of rods and chains from those rooms which the radiators to which they are attached supply respectively, but in this apparatus it is proposed to operate them by the automatic regulator hereinafter described.

The superiority of the low-pressure gravity-circulating apparatus has been completely established, so that no argument need be made for having adopted it in this design.

Systems of Piping.

There are several systems of piping for this class of apparatus, of which two are generally used for works of about the size of the one under consideration.

The principal features of these two are : For the first, main distributing-pipes and distributing-risers, and a separate return riser for each radiator or coil, which return risers are joined together below the water-line of the boiler.

For the second, main distributing-pipes and distributing-risers and corresponding return mains, but no return risers, the water of condensation returning through a relief at the foot of each riser.

The first of these systems has these advantages and has in consequence been adopted in this design :

1. With this system the apparatus can be made absolutely noiseless.

2. There is never any difficulty in expelling the air and it will circulate on an extremely low pressure.

3. A radiator which by mismanagement has been allowed to fill with water will empty itself noiselessly when the valves are opened.

Size of Steam-Mains

There is no part of a steam warming apparatus upon which its successful working depends more than on the size of the steam-mains and the manner in which they are run.

The rule given by Mr. Baldwin ("Steam Heating for Buildings"), "the area of a 1-inch pipe in the main at the boiler to each 100 square feet of radiating-surface, mains included," gives excellent proportions for apparatus of this kind.

Pitch of Pipe.

Steam-mains on leaving the boiler should at once be carried to the highest point, and from there their pitch should always be away from the boiler, causing the current of condensed water to flow in the same direction as the current of steam, making a relay where necessary to keep the pipe near the ceiling.

A main should not decrease in size according to the area of its branches, but very much slower, and the main to a riser at the end of a line should always be one size larger than the riser.

There is no need of a stop-valve in the steam-main near the boiler, because if necessary the steam can be run down in a few minutes by opening the feed-doors, &c., and damping the fire.

The ordinary practice in piping apparatus of the size of this one is to run the return mains and risers one size smaller than the steam. This makes them somewhat larger than is actually needed to carry back the water of condensation, but serves as a margin of safety against stoppage by dirt or sediment. Nothing smaller than 3⁄4-inch pipe should be run. *Return Mains and Risers.*

Return mains should all drip to a point near the boiler, their pitch being always in the direction of flow of the current of water. At the lowest point there should be a 1¼-inch steam-cock, with a 1¼-inch blow-off pipe connected to the cellar-drain outside the house-trap.

There should be an open-way valve in this return main between the blow-off and the boiler, by closing which the sediment in the pipes can all be blown out through the blow-off cock. No cluck-valve is required in the return main.

Where steam-mains are reduced eccentric fittings should be used to keep the bottom of the pipe on a line, otherwise pockets are formed which fill with water and are liable to cause snapping. *Eccentric Fittings.*

Mains should be run far enough from the walls up which risers go to give room for a riser connection about 2 feet long, to allow for expansion.

Tees in the steam-main for riser connections should look up, as by nippling an elbow on a riser connection can be made at any desired angle without straining the pipe. *Tees.*

Relief-pipes are necessary wherever traps are formed in the pipes, 3⁄4-inch being large enough for most places, except at the end of mains, where 1-inch should be used. *Relief-Pipes*

It is not necessary to discuss the merits of boilers of the "self-feeding" and "surface-burning" types for house-heating. The self-feeders are as economical of fuel and are structurally as nearly perfect as any other type, and have the advantage of keeping up steam for a much longer time without attention. *Types of Boiler.*

Upright
Tubular
Boiler. The boiler which has been selected for this apparatus (see Fig. 5) is built on the plan of an upright tubular boiler, with the addition of an arrangement for feeding the coal to the fire as it is consumed, from the sides. It consists primarily of two parts, a shell and a water-leg, both made of wrought-iron, with riveted joints, and connected to either by suitable circulating-pipes.

The shell is 41 inches in diameter at the top and 35 inches in diameter at the bottom and contains 84 tubes 2 inches in diameter and 34 inches long.

The water-leg is a hollow annular ring, with a section nearly triangular, and measures 45 inches in diameter outside and 32 inches inside, forming a fire-pot 14 inches deep, with sloping sides. The outside plate of this water-leg is carried up 14 inches above the water-chamber, and to its top edge is fitted the cast-iron ring which supports the shell, as shown. Where this plate extends above the water-chamber it is lined with iron plates bolted fast, and through it at equal distances are cut the holes around which the four coal-pockets are bolted.

Jacket. Around the upper part of the shell is a jacket of No. 20 galvanized iron, lined with asbestos sheeting ¼ inch thick, and stiffened at the top with a cast-iron ring.

The top of the boiler is covered with a removable cast-iron plate.

Base. The water-leg rests on a cast-iron base, forming an ash-pit 16 inches deep under the grate. This base has a double door at the front for removing the ashes, a small door on each side for access to the grate and a balanced draft-door for the automatic regulator.

There are fitted into the jacket, a little above the ring I, two soot-doors, a check-draft for the automatic regulator and a collar for the smoke-pipe.

Grate. The grate is made in two parts, an annular outside piece, consisting of a series of radial fingers joined to a ring, and an independent center-piece. The outside piece rests on balls, so it can be easily shaken by a lever with a handle at a convenient hight. The center-piece is arranged to swing one side, leaving a large opening, through which clinkers can be removed or the fire dumped.

This grate is 30 inches in diameter and sets about 3 inches below the bottom of the water-leg, as shown in sketch.

The space between the grate and the water-leg is occupied by a toothed ring, the object of which is to admit air to the fire when the grate is covered with an accumulation of ashes, enabling the fire to burn a long time without shaking. This feature is of importance to an "automatic" boiler, as the openings in the grate must be large enough to pass sufficient air when the fire is packed with ashes; in the last hour it is supposed to run without attendance.

The sloping sides of the fire-pot form a coal reservoir of sufficient capacity to maintain a steady fire and keep up steam from 12 to 24 hours without refilling, and the grate being low in the center, as the fire burns the mass of fire will move together from the sides centrally, as well as downward, thus keeping a compact mass of ignited coal on the grate for a long time. This is important when a boiler is expected to keep a fire for a long time without attention. A shallow fire soon burns out in places, but a deep fire formed as just described will keep in good condition for a long period. *Coal Reservoir.*

The form of "side-feed" boiler has the following advantages over those of the "center-feed" or "magazine" type : *Advantages.*

The coal in the reservoir is in the coolest part of the boiler, and does not, as in boilers of the "magazine" type, become highly heated, forming noxious gases, which escape through any openings into the house, and particularly when the cover of the magazine is removed to supply coal.

In feeding the coal has only to be lifted to the hight of the pockets, and not over the top of the boiler.

The weight of the coal in the reservoir does not rest on the grate, which, if it did so, would tend to warp the latter and make it hard to shake.

The fire is in the center of the grate, which is the most effective part for heating.

Access can be had to the fire from all sides for cleaning or kindling

The entire boiler is covered with an asbestos jacket 1 inch thick, to prevent radiation into the cellar.

A boiler for house-heating should always contain a quantity of water sufficiently large to fill the pipes and radiators with steam without lowering the water in the boiler enough to require an addition when steam is up. *Amount of Water.*

The radiators, mains, &c., of this apparatus have a cubical contents of about 30 cubic feet.

Taking 5 pounds as the maximum steam-pressure, 1 cubic foot of steam at that pressure weighs 1-20 pound. Allowing four times this amount for water in the steam, 8 pounds of water evaporated at this pressure would fill all the pipes and radiators and would lower the water-level less than ½ inch.

The fire-pot of this boiler is of iron, which prevents clinkers from forming at the sides and the necessity of repairs to brick-work, which are unavoidable with brick-lined furnaces. *Fire-Pot.*

Fire-
Chamber. The fire-chamber above the fire is spacious, which permits perfect combustion of the gases before they enter the tubes, which are large, and together with the other heating-surfaces are in a vertical position, so that they do not foul easily, and any deposit will fall off.

Besides having the heating-surfaces arranged to require the least possible cleaning, by removing the cover-plates from the top of this boiler easy access can be had to every part of it. This is important, as these boilers are often cared for by inexperienced persons (servants), who will condemn anything which gives them trouble.

Ash-Pit. The ash-pit is of good depth, with plenty of air-space under the grate as a precaution against its becoming over-heated, and is arranged so that a new grate can be quickly put in by any one.

Fire and Ash-
Pit Doors. The fire and ash-pit doors are planed to shut tight, and will completely damp the fire without closing a damper in the smoke-pipe. This method of damping the fire by shutting off the supply of air is the proper one, for the draft of the chimney being unimpaired draws all the harder at every crack or opening, causing an inward current at such points and preventing the escape of gas into the house.

The grate is made with radial arms, free to expand independently without causing internal strains and subsequent cracking.

Proportions
of Boiler. The proportions of this boiler are as follows

Heating-Surface.

	Square feet.
Shell	29.7
84 tubes, 2 inches diameter, 34 inches long	112.3
Tube-sheets, less area tubes	12.2
Water-leg	10.4
Total	164.6

Area of grate-surface	4.9 square feet.
Area of opening through tubes	1.5 square feet.
Area of smoke-pipe	0.44 square feet.
Ratio of grate-surface to boiler-surface	1 to 33.6
Ratio of tube area to grate-surface	1 to 3.26
Ratio of smoke-pipe area to grate-surface	1 to 11.1

Radiating-surface in building, direct	340½
" " " " indirect	410
Total	750½

Ratio of boiler-surface to radiating-surface	1 to 4.55
Ratio of grate-surface to radiating-surface	1 to 153

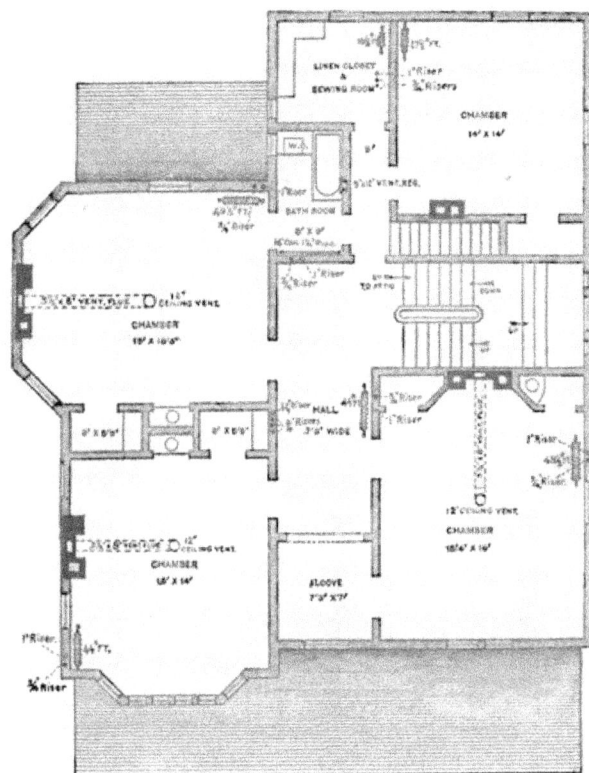

Fig. 3.—Plan of Second Floor.—Scale, 1-12 Inch to the Foot.

FLOOR PLANS ACCOMPANYING ESSAY OF ANSON W. BURCHARD.

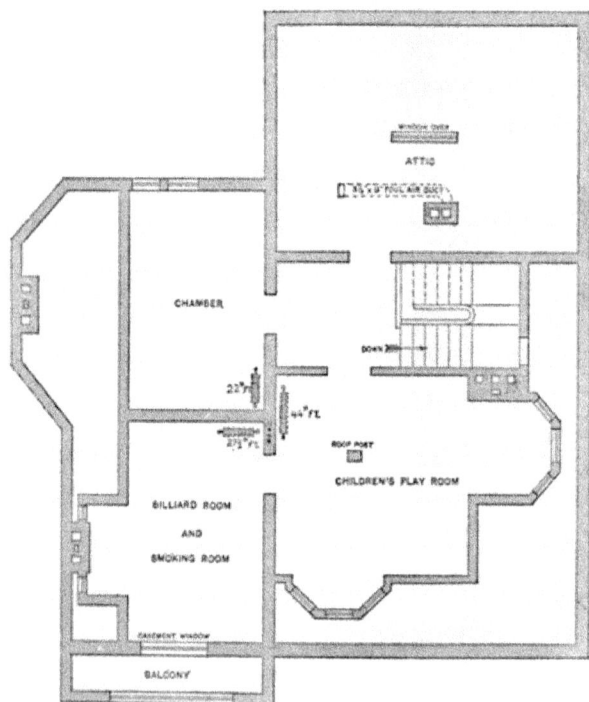

Fig. 4.—Plan of Attic.—Scale, 1-12 Inch to the Foot.

The condensation per square foot of average radiating-surface in a house of this class with an ordinary steam-pressure will not exceed ⅓ pound of steam per hour, including condensation in the mains. Per square foot of net surface it will average from 0.25 to 0.30 pound. This would give a total condensation of 250 pounds of steam per hour when all the radiators are turned on, calling for an evaporation of 1.52 pounds of water per square foot of boiler-surface per hour. There is no difficulty in getting an evaporation of 2 or 2½ pounds per square foot from these boilers. This large excess capacity is provided because the boiler will be called on to run for long periods without cleaning or interruption and be adequate to keep steam to every contingency of change within these limits of time.

A safety-valve is a necessary adjunct to every steam-boiler, whether for power or heating. In common practice no rule is used for determining the size of safety-valve required for a house-heating boiler, but the fitter or maker guesses at the size or puts on a certain size because he has been in the habit of doing so. Makers of boilers who send out trimmings with them generally send the same size safety-valve with every size boiler they make. The rules for determining the area of safety-valves, based simply upon the area of heating-surface, are generally adapted to high-pressure boilers with a high rate of combustion and are not applicable to house-heating boilers. Wilson ("Steam-Boilers") gives the following:

$$A = \frac{g\,c}{5.14\,P,}$$

where A equals the area of the safety-valve aperture, g the area of the fire-grate in square feet, c the maximum rate of combustion in pounds per square foot of grate per hour and P the maximum pressure in pounds. In calculating A the nominal area of the valve must not be taken, but the actual area of the aperture when the valve is raised from its seat, which latter seldom raises more than $\frac{1}{8}$ inch. Assuming the maximum rate of combustion to be 7 pounds per square foot grate-surface per hour and that the pressure is not to rise above 15 pounds in the boiler, we have:

$$A = \frac{4.9 \times 7}{5.14 \times 15} = 0.45 \text{ square inch,}$$

which is just the area of aperture of a 2-inch safety-valve with a conical seat when the valve is raised $\frac{1}{8}$ inch from its seat.

Condensation.

Safety-Valve.

The outlet to the safety-valve can be bushed to 1 inch and a pipe run from this into the ash-pit, so that in case the automatic regulator should fail to close the drafts when the steam-pressure rose the steam blown off through the safety-valve would fill the ash-pit and damp the fire.

Regulators. Of all kinds of automatic regulators for low-pressure steam-work the ordinary regulating-bowl with rubber diaphragm is the simplest and most reliable. It should be connected to the boiler with a ¾-inch pipe having a stop-valve and a water-syphon of ample capacity. The syphon protects the rubber from the action of the steam and the stop-valve enables a new diaphragm to be put in without letting the steam down, and serves to shut off the steam in case the rubber gives out. This connecting-pipe should be taken from the steam-space of the boiler, otherwise if the rubber gave out the water would be forced out of the boiler.

Steam-Gauge The steam-gauge should be attached with the usual syphon, and should be of a reliable make with anti-corrosive movement. The dial should be about 5 inches in diameter, with the figures plainly marked and indicating about 30 pounds.

Water-Feeders. Automatic water-feeders, although much favored, are not to be recommended. As yet a positive and reliable one under all circumstances has not been invented.

Water-Level The attendant to a boiler should look at the water-level once a day and if necessary let a little water in. This familiarizes him with the boiler, and he knows what to do in case anything happens.

Where an automatic feeder is used on a boiler, and, as they sometimes do, it fails to shut off the water and floods the boiler, the attendant is at a loss to know what to do and generally becomes much alarmed.

Water Column. The boiler should be fitted with a cast-iron water-column, connected, as shown, with 1-inch pipe, having a ½-inch sediment-cock at the bottom. This column should be fitted with three compression gauge-cocks and a glass water-gauge complete.

Steam-Cocks Asbestos-packed steam-cocks are the most reliable for sediment and blow-off cocks or for any place where water containing grit passes through them.

Radiating-Surface. To determine the exact amount of radiating-surface required to warm a given apartment is one of the most difficult—at the same time important—problems in proportioning a heating apparatus. There are a number of different ways of estimating this surface, but the conditions of exposure, tightness of windows, care with which a building is constructed, &c., are so variable that the results given by any rule must be care-

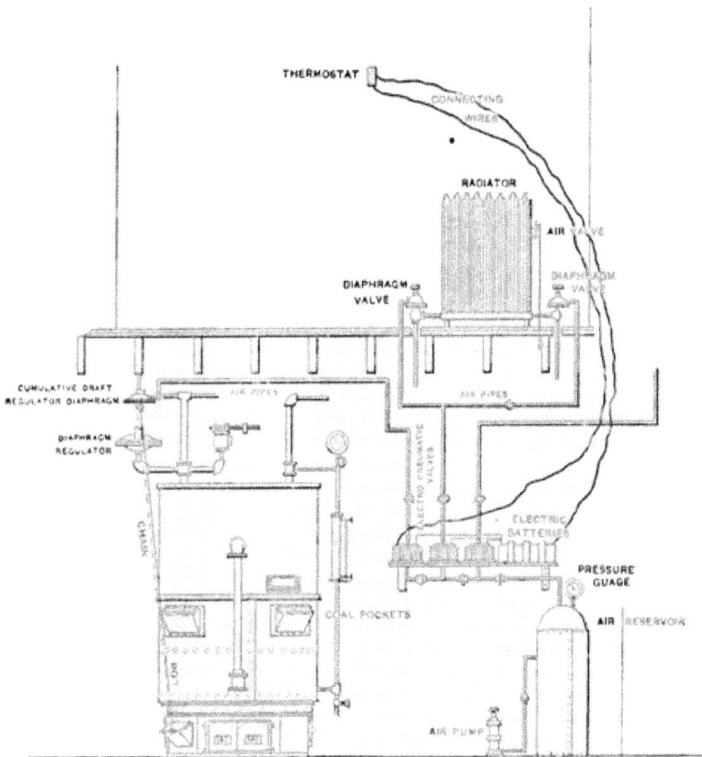

Fig. 5.—View of Boiler and Its Connections.

fully checked, taking into account these variations. As a room is cooled by the heat transmitted to the outer air through its windows and exposed walls, a rule for estimating the radiating-surface based on amounts of wall and glass surfaces exposed must approximate more closely than one which takes into account only cubical contents or glass surface alone.

For frame houses as ordinarily constructed 10 square feet of exposed wall surface will transmit about as much heat as 1 square foot of glass. Where the temperature of the room is required to be kept at 70° F., with the temperature of the outer air at 10° F., 1 square foot of direct radiating-surface to 2 square feet of exposed glass surface and its equivalent in wall surface (area glass exposed plus 1–10 area wall exposed) is required.

However, where the exposed surface is small in proportion to the cubical contents of the room, as 1 square foot to 75 or 100 cubic feet, a larger amount of radiating-surface than this rule calls for should be used, and in such rooms 1 square foot of radiating-surface to 60 or 80 cubic feet of space is required, because this amount of radiating-surface seems to be necessary in any apartment of moderate size heated by direct radiation to keep the walls at a temperature nearly that of the air by the direct rays of heat from the radiator, which adds very much to the comfort of the occupants of an apartment.

Bath-rooms should always be provided with a liberal amount of surface, because it is frequently desired to raise their temperature to a degree higher than that of the other rooms.

Indirect Radiation. For heating rooms by indirect radiation 1 square foot of radiating-surface is required to each square foot of exposed glass surface and its equivalent of wall surface. In proportioning the indirect radiators for the different rooms of the house under consideration it was thought best to take some surface from the radiators for the parlor, library and dining-room and add it to the radiator for the hall. This makes them and their registers all of about the same convenient size.

Indirect Radiators. The necessity for the use of indirect radiators in connection with the system of ventilation is obvious, and is a sufficient reason for their having been adopted in this apparatus. The indirect radiators which have been selected consist of common return-bend coils of 1-inch pipe, which pipes are tightly wound with spiral coils of No. 14 square wire, the whole being inclosed in a tight, substantial wooden box lined with tin.

This style of indirect radiator is light, compact, convenient to handle and put up, has no "packed" joints, allowing free circulation and at a low pressure. The main current of air through the coil is broken by the wire into a great number of smaller currents, so that a gust of wind cannot force the air through without its becoming heated, and its velocity is very nearly uniform at all times, not being affected by the fluctuations of the wind.

These radiators are supported by lugs bolted to the casing and hooked into ⅜-inch eye-bolts screwed into the joists.

Cast-iron direct radiators without bases or tops have ad- *Direct Radiators.* vantages which commend them for use in private houses, and have accordingly been adopted for this design. These radiators are made in loops or sections 2 or 3 inches wide by about 8 inches deep and generally 36 inches high. These sections are screwed together at the bottom by nipples, generally 2-inch, which form the connection for circulation. At the top the loops are fastened together by a long bolt.

These radiators are made by a number of manufacturers, the difference being mainly in the design. Those with no tops, rounded corners, planed edges and a figure in relief on the sides are the most ornamental. The legs should be detachable, to avoid danger of breakage in shipment, and the lower ends of the loops free from crevices or corners to collect dirt. One maker has eccentric bushings for the end sections into which the valves are screwed. By turning these bushings a slight error in the position of the valve outlets can be compensated for.

This style of radiator is very compact. Access can be had to all parts of it for dusting, and the floor under them can be easily cleaned, which cannot be done with radiators having bases.

Radiator-valves should be Jenkins' or Russells' own make *Radiator Valves.* soft-metal seat angle-valves, with ground joint-male union, finished bodies and plated all over. These valves besides having the well-known advantages of the soft metal disks are made heavy and with stuffing-boxes of good size. The finished bodies do not stain, and the male union permits a radiator to be disconnected with an ordinary monkey-wrench, while the ground joint avoids the necessity of using packing, which has to be renewed frequently. Spun brass nickel-plated floor-flanges with tubes 3 inches long are the best for the radiator connections.

Where radiators are set in carpeted rooms it is best to place under them rectangular boards about ½ inch thick long enough to take in both steam and return-pipes and painted to suit the wood-work of the room. With this arrangement the difficulties of fitting the carpet around the legs of the radiator and the connections are avoided.

Outlets.   Before a radiator can be filled with steam some means must be provided to allow the air to escape. If radiator-tubes were single with closed tops, when the steam was admitted the air would be forced to the top of the tubes in spite of the fact that air is about twice as heavy as steam. But if the radiator-tubes are in the form of a loop or with an internal diaphragm, when the steam enters the air falls to the base and is crowded to the end removed from the inlet, or will be forced out through an outlet if one is provided.

Air-Valves.   This outlet in common work is closed with a pet-cock or compression air-cock, but besides requiring attention when the steam is admitted to the radiator, these are objectionable on account of the water which is apt to be forced out with the air. There have been invented many forms of automatic air-valves for automatically closing this outlet when the air is out of the radiator, and all of which take advantage of the difference in expansion of two metals or materials having different ratios of expansion. These two materials are so arranged that when the steam comes in contact with and warms them the greater expansion of one than the other closes the outlet.

Those forms in which the piece forming the valve-seat has a certain amount of elasticity require the least adjustment and are the most durable.

To carry off a certain amont of water which is apt to be forced out with the air every automatic air-valve should be connected to a $\frac{1}{4}$-inch drip-pipe running to the cellar.

Large size, say $\frac{3}{8}$-inch, automatic air-valves with thumb-screw for convenient adjusting are the best to use on indirect radiators.

Cold-Air Currents.   The air of a heated room is cooled by contact with the windows and exposed walls, which causes a downward draft of cold air at such points; thus a circulation is established, the current of air passing along the ceiling, down on the exposed sides and in front of the windows, along the floor and up on the warm side of the room. This causes the " cold floors " of many imperfectly heated apartments.

These currents can to a large extent be neutralized by placing the heating-surfaces and introducing the heated air on the exposed sides and in front of the windows.

Position of Registers and Radiators.   In many large buildings having thick walls and recessed windows radiators are placed under the windows with excellent results, but in a frame building people object to this arrangement.

In this design radiators and registers are located so far as practicable in exposed parts of the room, taking into account the probable arrangement of the furniture, particularly in the bedrooms.

Return-bend coils of 1¼-inch pipe are used in place of radiators in the toilet-rooms where the floor space is limited.

The radiators on the attic floor are concentrated near the center of the house for convenience in piping. As these rooms are not much exposed, the results from the radiators so located would be satisfactory.

Radiators having 60 square feet or less of surface should have 1-inch steam valves and connections and ¾-inch return valves and connections, smaller than which should not be used for any size of radiator in an apparatus designed to circulate freely at a very low pressure. *Radiator Connections.*

Radiators larger than 60 feet, but less than 120, should have 1¼ and 1 inch steam and return valves and connections respectively. This rule applies to both direct and indirect radiators.

Open-way valves should be used for indirect radiators and elsewhere (except for direct radiators) in low-pressure steamwork. Although perhaps not quite so reliable as globe-valves, they answer every requirement for this work and offer the minimum resistance to the circulation. *Open-way Valves.*

The degree of relative humidity which air should possess for healthy respiration has been the subject of much inquiry and widely different views are held in regard to it. It is not probable that where a large volume of moderately-heated air is introduced into a room, as is proposed in this case, that its lower percentage of moisture would be sufficiently marked to cause discomfort, and efficient appliances for supplying moisture are so complicated or require so much attention that it can hardly be considered advisable to adopt any of them. *Humidity.*

A simple and fairly efficient way of supplying moisture is to set on each indirect radiator a dish of porous earthenware containing water. An opening must be made in the casing through which to replenish the water.

Registers with solid bronze faces are recommended for the indirect radiators. The first cost of these is somewhat greater than that of other kinds, but they will stand a great amount of wear without losing their good appearance. *Bronze-faced Registers.*

No floor borders are required, the registers being connected to the radiators with boxes of bright tin the full size of the register nailed tightly to the floor and the radiator casing. *Floor Borders.*

The fresh-air ducts are made of No. 24 galvanized iron of the dimensions marked on the plans and the pieces "double- *Fresh-Air Ducts.*

seamed" together where practicable, otherwise bolted with small screw-bolts. At the inlet end these air-ducts should be covered with galvanized wire-cloth, ½ inch mesh, to prevent foreign objects from getting into the pipes.

Strainers. The strainers consist of galvanized-iron frames, across each of which are stretched two sheets of brass wire-cloth, No. 30 wire, 1-16-inch mesh, with a space of ½ inch between the two sheets. The frame is arranged to be slid out for cleaning, which must be done when the filter becomes foul.

Dampers. A tight-fitting galvanized-iron damper is arranged in each duct, with a crank and rack in a convenient place for regulating the supply of air.

A substantial mixing-damper, also of galvanized iron, arranged as shown in Figs. 6 and 7 and having a suitable crank for attaching the automatic regulator, is provided in connection with each indirect radiator.

Position of Pipes. In general, pipes should be free to move in the direction of their length and not thrust against the parts of the building or lift or more radiators.

Expansion should be taken up by right-angle turns. They should be perfect in their alignment and the pitch should be in the direction of flow of the current of condensed water or steam, as the case may be.

Horizontal steam-mains should be supported at intervals of 10 feet by iron hangers screwed into the joists.

Return-mains are best run under the cellar-bottom before it is cemented and tested and made tight before covered over.

This leaves the cellar-bottom free from obstruction and the chances of having to dig up the pipes are too remote to be considered.

Right and Left Couplings. All connections should be made with right and left threaded couplings.

Slip-Tubes. Where pipes pass through partitions, walls or floors they should be protected with galvanized-iron slip-tubes with cast-iron flanges at their ends. These tubes should be seamed together and slipped over the pipes.

Painting. Direct radiators, after being connected, should be painted with a smooth mixture of yellow ocher, white lead and boiled linseed-oil, which should be steam-baked after it is applied and then hardened by air-drying. They can be bronzed in gold or silver, as may be desired.

The pipes in the cellar and the iron-work about the boiler should be painted with two coats of black air-drying japan.

Removable Pipe-Covering. All the steam-mains and their fittings should be neatly covered with removable pipe-covering composed of a layer of asbestos paper on the inside, then a layer of asbestos fiber, another layer of paper, a layer of hair felt and an outside cover of tough paper, the whole making a thickness of about 1 inch.

In this design provision is made for carrying off the depreciated air, principally through the open fire-places, of which there are six.

The parlor, library and dining-room are also furnished Ceiling Ventilators. with ceiling ventilators connected with separate flues, the intention being to carry off the products of combustion of the gas without allowing them to mix with the air of the room to any great extent. The capacity of these flues is hardly large enough to carry off the estimated supply of air, but larger flues could not be built into the chimneys without making radical changes in their dimensions. All flues in the chimneys should be lined with smooth pottery linings.

The smoke-flue for the boiler should be carried below the Smoke-Flue. point where the smoke-pipe enters it and at the bottom a cast-iron door-frame and door built into the brick-work to permit access for cleaning the flue.

The bath and toilet rooms are provided with ventilating- Bath-Room Ventilating-registers connected by tin pipes with a separate flue in the Registers. kitchen chimney. As this chimney is always warm this flue will be constantly drawing air from the other rooms into these rooms, preventing the dissemination of unpleasant odors.

AUTOMATIC DRAFT-REGULATOR.

It has already been stated that perfect ventilation may be said to have been secured in an inhabited room only when it has been established in connection with the perfect comfort of the occupants as regards temperature, being neither too hot nor too cold.

With such an apparatus as has been described these results would not be obtained unless the steam-pressure be controlled by other means than the ordinary damper-regulator and the supply of heat to the room by other means than the registers and ordinary radiator-valves.

In order to shut off the heat from a radiator it is necessary to close the valves entirely. If the room gets too warm it is essential to close the valves tightly; in this case the room soon becomes too cold. It is then necessary to turn the steam on again, and thus unless the pressure of steam is accurately adjusted to the room, radiating-surface and weather the valves require constant attention. The radiator-valves should both be either wide open or tightly closed ; for if turned only a short way to "let a litte heat in" the condensed water will not run off and the radiator will fill with water. Through lack of knowledge or carelessness it often happens that one valve will be left open more or less, with the other one closed. Valves and registers are often closed and forgotten, so that the room is found cold when it should be warm.

As it is impossible to so proportion and arrange a heating apparatus as to maintain a uniform temperature in all weather without constant attention, some means must be taken to automatically control the supply of heat to each room. For this purpose many automatic regulators have been invented, some of which have been perfected to such a point that when properly applied they will maintain the temperature within one or two degrees of the standard, unless, of course, the fire be allowed to go out.

The apparatus which has been adopted in this design will be readily understood from the detail drawing.

The peculiar feature of this system is the electro-pneumatic valve—that is, a pneumatic valve operated by electricity.

The motive power for opening and closing the radiator-valves and for operating the dampers of the indirect radiators is compressed air, which is stored in an iron tank. This tank is provided with a pressure-gauge and a small hand-pump with which the attendant keeps the air-pressure in the reservoir up to the required point. This requires only a minute or two of pumping daily, but if desired an automatic pump operated by water under pressure from the service can be used instead.

In each room of which it is desired to control the temperature, or in several of the principal rooms where there are a number of rooms always open, is placed a thermostat, which is connected by electric wires to a corresponding electro-pneumatic valve.

Where rooms are heated by direct radiation the radiators are fitted with "diaphragm-valves" in place of the ordinary radiator-valves. The bodies and disks of these valves are similar to those of the ordinary radiator-valves, but the stem is without a screw and passes through a stuffing-box into a chamber, where it is attached to a piston. Across this chamber is stretched a rubber diaphragm and on top of this is bolted a cover-piece, forming between it and the rubber a tight chamber. When the air under pressure is admitted to this chamber the diaphragm pushes the piston down and forces the valve disk on to its seat. If the air is allowed to escape from the chamber a spring opens the valve to its full extent.

Each indirect radiator is fitted with a mixing-damper, already described, to the rod of which is fastened a small pinion.

Fastened to each air-duct is a small diaphragm apparatus, which moves a rack, which in turn engages the pinion and turns the damper.

Each diaphragm-valve and diaphragm apparatus is connected to its respective electro-pneumatic valve by a small composition tube and each valve to the reservoir in the same manner. Suitable electric batteries are provided to furnish the requisite electric current.

The action of these appliances is as follows: When the apartment reaches the temperature at which the thermostat is set the electric

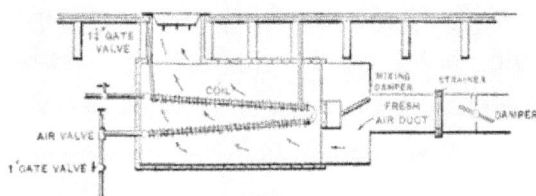

Fig. 6.—Section of Indirect Radiator and Casing, Showing Mixing-Damper.

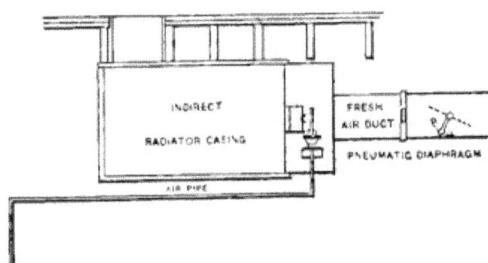

Fig. 7 —Side Elevation of Indirect Radiator, Showing Method of Attaching Pneumatic Diaphragm to Mixing-Damper.

circuit is closed ; this operates the electro-pneumatic valve, which allows the compressed air to enter the diaphragm-valve and shut off the steam. Of course the temperature soon begins to fall. When it has fallen a degree or so the electric circuit is again made and the electro-pneumatic valve allows the air to escape from the diaphragm-valves, which open and admit the steam to the radiators.

With the indirect radiators the action is similar. When the circuit is made the pneumatic valve admits the air under pressure to the diaphragm, which moves the pinion and causes the mixing-damper to assume a position such that the air will pass over instead of through the coil. When the temperature falls and the circuit is again made the air escapes from the diaphragm and the damper assumes the first position.

With ordinary valves it always gets too warm before the heat is noticed and the steam turned off or the register closed, and then it gets too cold before the valves are thought of. But with this apparatus, since the supply of heat is shut off as soon as the room is warm enough, a room never gets too warm; and if sufficient steam is provided never too cold.

The valves are always fully opened or fully closed, and are open when steam is down. The electric valve shuts off the radiating-surface in the rooms that are warm enough and the steam-pressure rises accordingly, which distributes the heat to the other rooms, so that enough pressure can be carried to warm the coldest room in a building without overheating any other room.

An electrically actuated diaphragm is attached to the same dampers to which the ordinary steam-draft regulator is attached without in the least interfering with the action of the latter.

This is arranged so that when the heat is shut off from all the rooms the damper admitting air under the grate is closed and the check-draft opened, whether the steam is up or not. Double thermostats are provided for the living-rooms, whereby a lower temperature may be kept at night if so desired.

Room.	Hight in feet.	Cubic contents, in cubic feet.	Number of windows.	Area of glass exposed, in square feet.	Area of walls exposed, in square feet.	Equivalent of walls in glass.	Total glass exposed, in square feet.	Radiating-surface direct. Square feet.	Radiating-surface indirect. Square feet.	No. cubic feet space to 1 square foot radiating-surface.	Size register admission.	Area ventilating-flue. Square inches.
Parlor	10½	3,045	4	84	273	27	111	....	100	30	14 x 22	64 x 28
Library	10½	2,835	8	112	245	24	136	....	110	25	14 x 24	64 x 28
Lower hall	10½	2,625	.	....	....	....	....	....	100	26	14 x 22	....
Dining-room	10½	2,919	4	82	298	30	112	....	100	29	14 x 22	64 x 28
Toilet-room	10½	189	1	....	....	....	....	8	....	24	......	31½
Second floor:												
North chamber	9½	2,327	3	48	261	26	74	49½	....	46	......	64
Front chamber	9½	2,125	7	72	213	21	93	44	....	48	......	64
West chamber	9½	2,641	3	54	298	30	84	49½	....	53	......	64
Upper hall	9½	2,424	4	56	....	....	56	44	....	55	......	....
Bath	9½	427	1	....	....	....	....	8	....	53	......	31½
Sewing-room	8½	571	2	....	....	....	....	16½	....	34	......	......
Rear chamber	8½	1,666	2	24	216	21	45	27½	....	61	......	......
Third floor :												
Play-room	9	2,394	6	24	....	....	24	44	....	54	......	......
Billiard-room	9	1,629	4	26	....	....	26	27½	....	60	......	......
Chamber	9	1,485	2	12	....	....	12	22	....	67	......	......
Totals	...	32,232	...	...	...	...	...	340½	410	..	..	....

Total cubic contents of rooms heated, 32,232 feet. First floor, 14,037. Second floor, 12,687. Third floor, 5508.

Total radiating-surface direct, 340½ square feet. Total radiating-surface indirect, 410 square feet. Total radiating-surface, 750½ square feet. First floor, 418 square feet. Second floor, 239 square feet. Third floor, 93½ square feet.

Cubic feet of space to 1 square foot radiating-surface : First floor, 33¼. Second floor, 53. Third floor, 59½.

Area of openings of registers : Total, 840 square inches.

Area of section fresh-air ducts : Total, 730 square inches.

Area of section of ventilating-flues : Total, 521 square inches.

Air warmed per square foot grate-surface in boiler, 6578 cubic feet.

Table Giving Dimensions of the Principal Parts of this Apparatus.

ESTIMATE OF THE COST OF THIS APPARATUS.

1 bolier, 30-inch fire-pot, 41-inch shell	$252.00
Asbestos jacket	20.00
340½ square feet direct radiators, at 30 cents	102.15
410 square feet indirect radiators, cased, at 26 cents	106.10
3 14 x 22 registers with bronze faces; 1 14 x 24 register with bronze face	67 00
2 9 x 12 electro-plate registers	6.00
3 ceiling ventilators	15.00
11 1-inch N. P. W. H. Jenkins radiator-valves, finished bodies, male union; 11 ¾-inch N. P. W. H. Jenkins radiator-valves, finished bodies, male union	51.80
11 N. P. automatic air-valves for drip-pipe	11.00
11 1-inch N. P. brass floor-thimbles; 11 ¾-inch brass floor-thimbles	3.30
1 6-inch steam-gauge; 1 inch safety-valve; 1 9-inch regulating-bowl; 1 water-column complete	18.00
1 2-inch open-way valve	3.25
1 1¼ asbestos-packed cock	2.75
2 ½-inch lever-handle stop-cocks	1.20
1 ¾-inch Jenkins globe-valve	.90
4 yards chain	.40
4 1¼-inch open-way valves; 4 1-inch open-way valves	12.00
4 ⅜-inch automatic air-valves	3.00
200 pounds galvanized-iron air-duct put up, at 20 cents	40.00
2 square feet wire-cloth, ½-inch mesh	.40
4 dampers	$1.00
4 rods, cranks and racks	8.00
4 Mixing-dampers	4.00
4 strainer-frames	6.00
16 square feet fine brass wire-cloth	2.40
4 sets hangers for indirect radiators	6.00
Register-boxes: 3 14 x 22; 1 14 x 24; 2 9 x 12	4.00
55 feet tin pipe 3½ x 9; 28 feet tin pipe 3½ x 8	20.75
30 pounds galvanized smoke-pipe	2.10
Bronzing 11 radiators	13.75
Painting pipes	6.00
Removable covering: 13 feet 2½-inch; 17 feet 2-inch; 32 feet 1½-inch; 33 feet ¼-inch; 40 feet 1-inch	25.50
Red lead, oil, waste, screws, nails, &c	5.00
Wrought-iron pipe: 13 feet 2½-inch; 35 feet 2-inch; 70 feet 1½-inch; 175 feet 1¼-inch; 230 feet 1-inch; 303 feet ¾-inch; 40 feet ⅜-inch	40.00
Cast-iron fittings	23.00
Labor, 27 days, man and helper.	135.00
Contingencies	15.00
Total	$1,033.75
Automatic draft-regulator complete	275.00
Total cost	$1,308.75

III. HOT-WATER CIRCULATION.

HOT-WATER CIRCULATION.*

BY RICHARD SWALWELL.

ADVANTAGES AND GENERAL FEATURES.

After an experience of many years in the planning and fitting up of heating apparatus, the writer is of the opinion that hot-water circulation is the best known medium for heating buildings, whether large or small. Whether situated in a climate where the thermometer seldom registers zero, or in a climate where the mercury for months seldom rises above zero and where 45° below is not uncommon, hot-water circulation, in the writer's experience, has given and is giving satisfactory results, both as regards efficiency and economy. As a system of heating, hot-water circulation possesses advantages which recommend it to all interested in heating apparatus. It is easily managed. Ordinary help is all-sufficient. It is perfectly noiseless. No pounding or hissing sounds. It produces a steady, mild, equable temperature throughout the building. No overheated or burnt air. It is economic. The circulation starting as soon as the fire is lit, transmits heat at once to radiators, and it can be easily regulated to suit the weather. It is healthy. No coal gas, dust or vitiated air. With ordinary care repairs are *nil*, while complete command is obtainable over all or any portion of the apparatus, any desired reasonable temperature—with a well-proportioned apparatus—being easily obtained. In the writer's opinion, the direct system of heating is the best for the following reasons: The occupants of the different rooms have, with the direct system, easy access to and command over the heat supply of their respective radiators, while with the indirect system it is difficult to adjust the register valves to obtain the same result. Again, should an indirect radiator or coil be turned off by closing the valves and the water it contains not be drawn off, there is the danger of freezing, owing to the cold air having access to the inclosure in which it is placed.

The indirect system is not as clean or healthy as the direct, as matter is liable to accumulate in the ducts and register-boxes during the summer, to be returned to the rooms during the winter months, vitiating the air supply. The radiators should be of cast iron, as this material affords a more pleasing form of radiator than box-coils of wrought-iron pipe. Again, though box-coils are, in the writer's opinion, rather more effective foot for foot, as commonly rated, in radiating power than cast-iron radiators, yet box-coils are open to the objection that

---

* From *The Metal Worker*, March 23, 1889.

uust, fluffy matter, &c., accumulate on the upper side of the pipes, and in a comparatively short time seriously interfere with their heating capacity. This is especially the case where box-coils are covered with ornamental screens and marble slabs. The writer has on several occasions examined heating apparatus that had become in-effective, and found the upper side of pipes coated with an accumu-lation ½ inch to ¾ inch thick which resembled in texture and appear-ance a loose felt. This matter proved a good non-conductor, as after its removal the apparatus worked efficiently. For these reasons the writer prefers using cast-iron radiators, preferring those that have the least receptacle for accumulating dust and that are easily accessible in all parts, allowing of daily cleaning or dusting, as with any other article of house furniture. Wall coils of wrought-iron pipe are not open to the above objection, and where they can be suitably used are cheaper than cast radiators, and as they consist of a single row of ex-posed pipes, can be easily kept clean. Wall coils when neatly set up on ornamental plates and nicely bronzed look fairly well, and it has been the writer's practice to use them on upper floors and basements where suitable wall space could be obtained.

In specifying the boiler and radiators, I will not describe any par-ticular boiler or radiator, simply confining myself to naming those points which I consider a good hot-water boiler or radiator should possess. In the general placing of apparatus I will be guided by that arrangement, both as regards sizes of mains, placing of radiators, heat-ing surface, &c., that many years' experience in the placing of hot-water apparatus has proved to be effective and sure to do the work called for. The system of ventilation to be described is one that, in the writer's practice, has given satisfactory results, and is original—so far as his knowledge goes—in that portion which gives each floor of the building a separate fresh cold-air duct and air-warming chamber.

The boiler is shown on basement plan, Fig. 1, set in hall or cor-ridor. This plan I consider best for the following reasons: The boiler is close to fuel supply or coal cellar. The most direct run of mains is secured in this position. Being exposed and easy of access, it is more liable to inspection by the householder, while in nine cases out of ten the help is sure to keep boiler and surroundings cleaner than if set in a less exposed place. Being a portable boiler it occu-pies very little floor space, say 3 x 3 feet over all, and when set as shown is an active agent in ventilating basement. The boiler grate, with its shaking attachment, should be so constructed that it can be easily shaken with the ash-pit or draft door closed tight, so that ashes and dust may not escape.

It will be observed, by referring to plans, that radiators and warm-air ducts are placed, as far as possible, in such manner that a thor-ough circulation of the air in the rooms is secured. The air-warming chamber consists of a frame structure of 4 x 2 studding. In the par-tition walls set the studding so that the smallest, or 2-inch thickness, makes the depth of wall between compartments. When the frame is

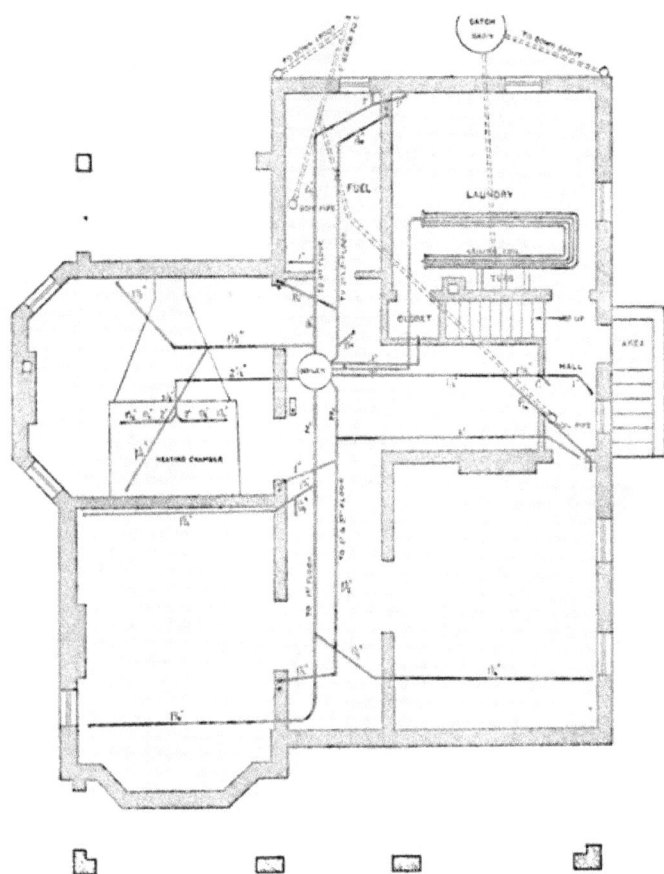

Fig. 1.—Cellar Plan, Showing Hot-Water Pipes.—Scale, 1-12 Inch to the Foot.

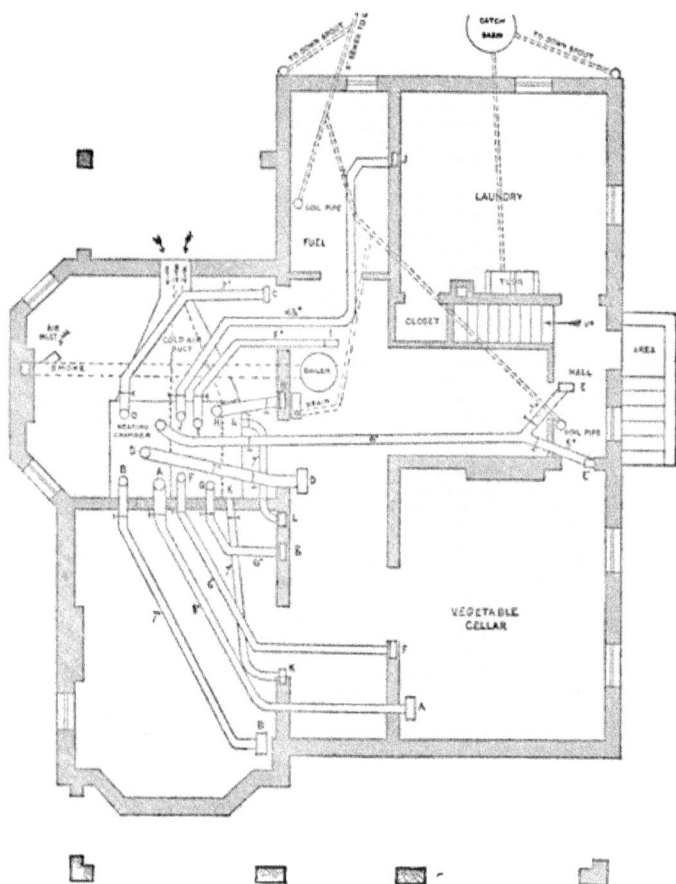

Fig. 2.—Cellar Plan, Showing Hot-Water Pipes.—Scale, 1-12 Inch to the Foot.

up let the tinner make three tin boxes the size of the different compartments, 28 inches deep. These being inverted, the collars for the different warm-air ducts may be fitted in, when they can be set up in place and nailed to studding. The coils are then placed and pieces of 1¼ pipe screwed into headers long enough to project 2 inches beyond finished outside wall of chamber, care to be taken that these pipes have the proper fall. When this is done the carpenter may line the compartments from floor up with tin. The tin at this point to have an edge 1 inch wide, bent inward ; this is well nailed to top board. Boards to be tongued and grooved. The top can then be covered with boards as desired. Sides and ends may be lathed and plastered or boarded. This chamber makes quite a safe and a much cleaner job than a brick chamber.

Traps or man-hole doors should be placed allowing of ingress to the compartments for annual inspection and cleaning, if necessary. In estimating, the figures given are those ruling where I am located. In the sketches inclosed I have tried to still further make my meaning clear.

SPECIFICATION.

The boiler, Fig. 6, to be a portable hot-water boiler, easily **Boiler.** accessible in all the parts exposed to the fire and gases of combustion, so that it can be easily and effectively cleaned when necessary.

Area of grate surface to be not less than 750 square inches. **Area of Grate** Area of effective heating surface—*i.e.*, surface fully exposed to **and Heating Surface.** the action of fire and gases of combustion—not less than 80 square feet.

Should construction of boiler render necessary, it should **Non-conducting Covering.** be covered with 1½-inch coating of asbestos or other good non-conducting cement.

The boiler to be set where shown on plan, Fig. 1, on a **Boiler Foundation.** solid brick foundation laid perfectly level and even with floor of basement. Make foundation 4 inches wider than boiler base, and 2 feet 6 inches longer at ash-pit door end.

Smoke-pipe to be No. 22 galvanized iron, 9 inches diameter, **Smoke-Pipe** with air inlet damper, Fig. 9.

At lowest point of boiler fit in a 1-inch brass draw-off cock, **Boiler Draw-off.** to empty boiler and apparatus when necessary. From this cock carry a 1-inch malleable-iron pipe to receiver.

The main flow and return pipes to be of the size marked **Mains.** on plan, Fig. 1. The return-pipes are not shown. Branches from same to be taken off in manner shown on plan by proper reducing fittings. The main-pipes to first floor radiators to be separate, as shown, and in no way connected with radiators on second or third floors.

Fittings and Elbows. Y fittings to be used rather than ⊺'s for branches, and 45° elbows or set-pipes to be used rather than right-angled elbows, where practicable.

Main Shut-off Valves. On all main flow and return pipes, and close to where they connect with boiler, fit on heavy brass "Peet" or "gate" valves to shut off the water and regulate supply.

Grading Mains. All horizontal mains to have a rise from the boiler to risers of not less than ⅛ inch to the foot.

Connections with Heating Chamber. The main flow to heating chamber to be laid as shown in sketch, Fig. 6, falling toward heating chamber, with air cock at highest point over boiler. Main flow 2½-inch pipe. The return-pipes from heating chamber to fall toward boiler. Main return 2½-inch pipe. The connections from mains to heating chamber coils to be carried out, as shown in sketch, with "Peet" valves, draw-off and air cock to each coil, as shown.

Risers. Main risers to upper floors to be carried up in partition where shown, Fig. 1, and branches from them laid to points where radiators are located on plan.

Testing Pipes Before lathing is placed or floors laid, plug openings and test with force pump 40 pounds water pressure. Securely cap or plug all open ends, and leave until ready for connecting to radiator.

Right and Left Couplings. All main-pipes to have right and left couplings near valves, on the side furthest from boiler, and all branch mains where they leave main-pipes to have right and left couplings, so that should a valve need replacing, or any alterations be made at any time, pipes may be easily detached.

Pipe Supports. All mains to be supported on wrought-iron hangers, firmly fastened to joists and placed at intervals of 10 feet.

Draw-off Valves and Pipes. On that side of valves furthest from boiler tap each flow and return main, and fit on a ⅜-inch brass draw-off cock, connected with ⅜-inch pipe to receiver, so that should it be necessary from any cause to empty a set of mains and radiators they may be emptied without interfering with the working of the remainder of apparatus.

Pipe Coverings. The branch main to parlor radiators can be laid between joists over vegetable cellar ; if this is not practicable cover with asbestos cement or hair felt, ¾ inch, if necessary. Mains running through fuel cellar to be covered with asbestos cement or hair felt.

Receiver for Draw-off Pipes. A tinned copper receiver or sink to be placed near boiler, where shown on plan, Fig. 2, with 2-inch trapped waste connected to drain, as shown in sketch, to receive water from all draw-off pipes from mains and boiler.

Waste-Pipe Stop-Cock. The waste-pipe from receiver to have an iron stop-cock, full bore of pipe, immediately under or near receiver, to be kept closed when not in use. This as a precaution against

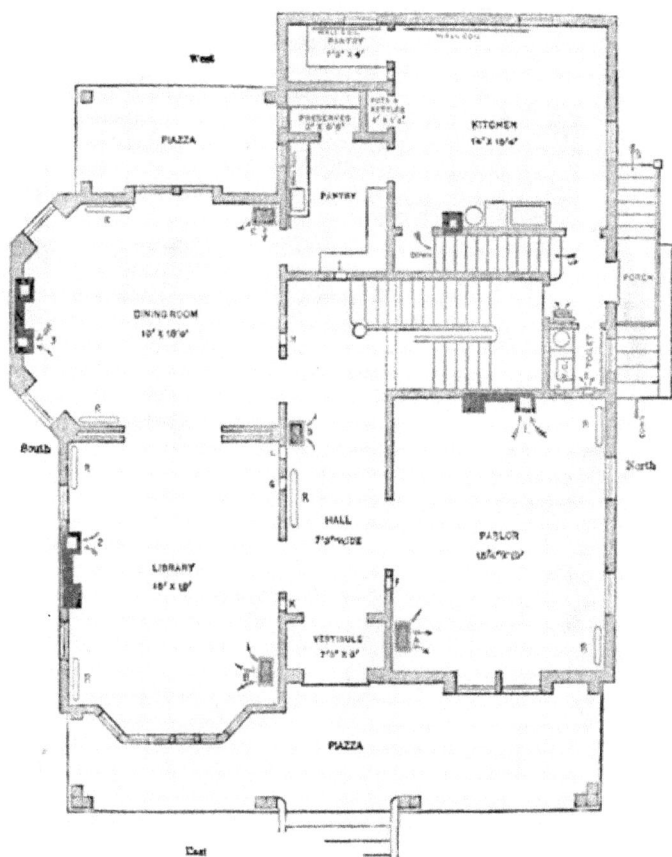

Fig. 3.—First-Floor Plan.—Scale, 1-12 Inch to the Foot.

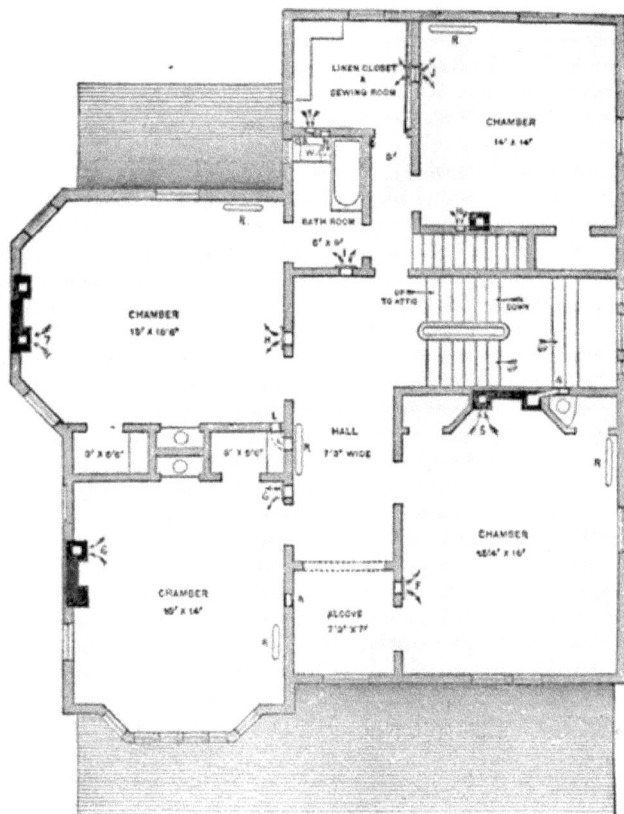

Fig. 4.—Second-Floor Plan.—Scale, 1-12 Inch to the Foot.

sewer gas, as receiver would be rarely used oftener than once
or twice a year, and not at all in the summer months.

Set up in attic, close to house-water supply-tank, an
expansion tank, Fig. 7, made of No. 24 galvanized iron, 28
inches high, 14 inches diameter, with brass mountings and
glass gauge. Seams and ends riveted and soldered tight,
and support it on strong wood or iron brackets, setting supports at such hight that bottom of tank will be 9 inches
below high-water level in house-water supply-tank. *Expansion Tank.*

From bottom of expansion tank carry a 1-inch pipe in
as direct a way as possible to boiler, and connect to inlet at
bottom of boiler. No valve or stop-cock to be placed on this
pipe under any consideration. *Expansion-Pipe.*

From house-supply tank lay a 1-inch pipe with stop and
check valve and connect to expansion-pipe under expansion
tank. The cock in feed-pipe being kept open and expansion
tank set as directed, an automatic and reliable feed is
secured, the check valve preventing return of water to
house-supply tank when water is lowered in it. *Feed-Pipe.*

In side of expansion tank, say half-way up, fit in a
1-inch brass flange, screw in a 1-inch nipple and elbow
turned up; into elbow screw a 1-inch brass cock (cock not
shown in sketch) with funnel on upper end. This feed is
put on in case of failure of water supply, is cheaper than a
pump, and from the small quantity of water needed to run
apparatus when once filled is all that is necessary. *Auxiliary Feed.*

From top of expansion tank carry a 1-inch pipe and
drop over nearest sink. *Overflow.*

From top of expansion tank carry a 1¼-inch pipe to under
roof, Fig. 8, and there enlarge to 2 inches, and carry 18 inches
above or through roof. On top of 2-inch pipe screw on a
galvanized-iron coupling. Cut opening in roof 5 inches in
diameter. *Air-Pipe.*

Make tapered galvanized-iron casing, No. 26 gauge, long
enough to reach from top of air pipe to roof. Flash roof securely round opening, and solder tapered casing firmly to
galvanized coupling on top of air pipe, and at bottom make
secure and tight to roof. This casing allows the warm air
from attic to surround air pipe to extreme end, preventing
condensation and possible freezing, an occurrence by no
means rare. *Casing Air-Pipe.*

Each radiator and coil to have heavy pattern brass angle
valve on flow or feed connections, and a ¼-inch brass air-cock
fitted on at highest point. *Radiator Valves and Air Cocks*

The wall coils to be set on wall above baseboards, so as
not to interfere with cleaning of baseboards or sweeping, &c.
Wall coils to be secured to wall with fancy cast-iron wall
plates, backed with wood and firmly screwed to studding. *Wall Coils.*

Wall Plates. Wall plates should be so constructed as to give a slight incline to pipes, so that air may freely rise to air-cock on upper pipe.

Radiators. Cast-iron radiators to be set where shown on plans, and to contain the number of square feet of heating surface specified below:

Radiators 1st Floor: Ft. heating surface.
  Parlor, 2 radiators, each containing.............. .. 45 = 90
  Library, 2 radiators, each containing.............. 35 = 70
  Dining-room, 2 radiators, each containing.......... 40 = 80
  Main hall, 1 radiator, containing.................. 72 = 72
  North passage, 1 radiator, containing.............. 32 = 32
Radiators, 2d Floor:
  Chamber, northeast, 1 radiator, containing.... ..... 48 = 48
  Chamber, northwest, 1 radiator, containing......... 56 = 56
  Chamber, southeast, 1 radiator, containing......... 40 = 40
  Chamber, southwest, 1 radiator, containing......... 65 = 65
  Hall, 1 radiator, containing....................... 56 = 56
Radiator, 3d Floor:
  Playroom, 1 radiator, containing................. .... 56 = 56
      Total feet........... 665

Wall Coils, 1st Floor:
  Toilet, 1 wall coil, containing........... 40 lineal feet 1-inch pipe.
  Kitchen, 1 wall coil, containing..........100 " " " "
  Pantry, 1 wall coil, containing........... 30 " " " "
  Pantry (room under sink), 1 wall coil, containing............................. 50 " " " "
Wall Coils, 2d Floor:
  Bathroom, 1 wall coil, containing........ 60 " " " "
  Linen closet, 1 wall coil, containing...... 50 " " " "
Wall Coils, 3d Floor:
  Billiard-room, 1 wall coil, containing..... 80 " " " "
  Chamber, 1 wall coil, containing......... 70 " " " "
  Attic, 1 wall coil, containing ....100 " " " "
Basement Coil:
  Laundry, 1 wall coil, containing..........100 " " " "
      Total........... ....680 " " " "

Smoke-Pipe. The smoke-pipe to be of material and size before specified, and to be carried direct as possible to flue in S. W. room of basement.

Air Inlet Draft Check. In the smoke-pipe and close to flue take out a 7-inch Y-branch and fit a 7-inch diameter valved register to check draft. This damper, with draft door on boiler, is all that is necessary to check or regulate fire. To this register attach a weight, chain pulley, &c., and carry chain up through floor to dining-room at side of chimney, so that, if necessary, it may be opened or closed from that point, as well as

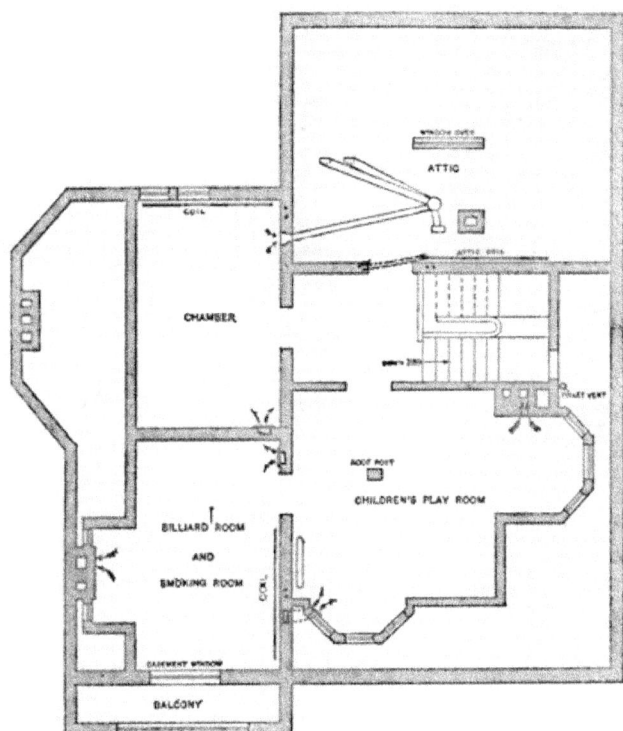

Fig. 5.—Attic Plan.—Scale, 1-12 Inch to the Foot.

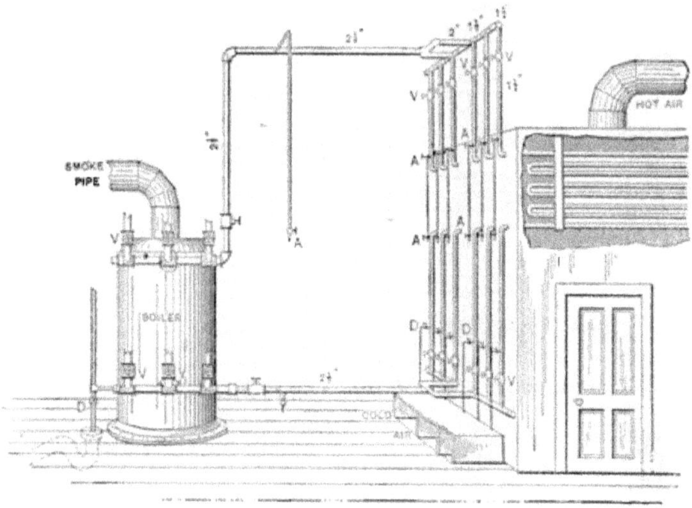

Fig. 6.—Boiler, with Hot-Air Chamber.

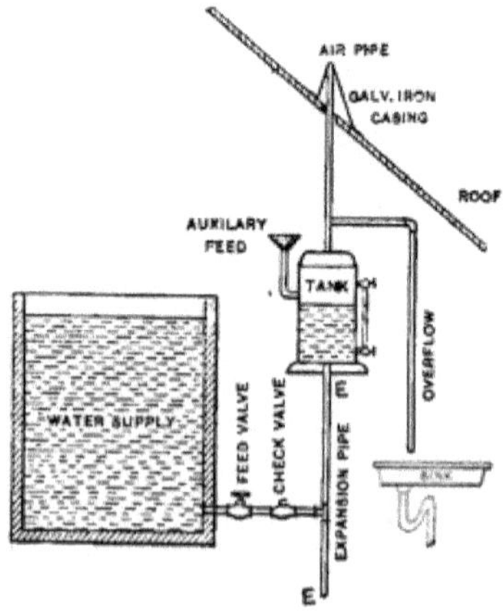

Fig. 7.—Expansion Tank.

from basement (see Fig. 9). Weight and chain pulley in dining-room nickel-plated.

Radiators on first floor (Fig. 3) to have 1¼-inch connections, excepting radiator in north passage, to which make 1-inch connections. *Size of Connections.*

Radiators on second and third floors, 1-inch connections.

Wall coils to have 1-inch connections throughout.

Risers to S. E. chamber, 1½-inch; thence up to billiard-room and playroom, 1¼-inch.

Risers to S. W. chamber, 1¼-inch; thence up to chamber, third floor, 1-inch.

Risers to N. E. chamber, 1-inch.

Risers to N. W. chamber and linen-room, 1¼-inch.

Risers to hall, second floor, 1-inch.

Risers to bath and attic, 1¼-inch; to second floor to attic, 1-inch.

Branches from risers for upper floors use return-bend fittings, with end opening to main flow 1¼-inch in diameter. The two upper outlets are 1-inch each, one of which leads to radiator and the other to riser. *Fittings for Riser Branches.*

Fit up close to laundry ceiling and hang from joists, on iron hangers, one ceiling circulation, shown in Fig. 1, with separate 1-inch flow and return pipe, valved at boiler only. At highest point fit on ¾-inch pipe and carry down over tubs and fit on air-cock. *Laundry Coil*

On plan (Fig. 1), flow-pipes only are shown, to avoid complication, the return-pipes to be of same size with connections same as flow-pipes. *Size of Flow and Return Pipes.*

Smoke-pipe flue about 10 x 8 inches, not less than 8 x 8 inches, carried up straight as possible. *Smoke-Pipe Flue.*

Fresh-Air Supply.

In southwest room, basement plan (Fig. 2), build a chamber with 4 x 2-inch studding for frame 8 feet long x 5 feet 3 inches, inside measure, and 6 feet 9 inches high from basement floor to top. Divide this chamber from top to bottom into three compartments, Nos. 1, 2, 3, as in Fig. 10. Compartment No. 1, 15 inches wide x 5 feet 3 inches; compartment No. 2, 30 inches x 5 feet 3 inches; compartment No. 3, 45 inches x 5 feet 3 inches. Line top of each compartment with tin and carry lining down, covering sides and ends, to a point 28 inches from top of chamber. In compartment No. 1 set up a box coil 5 feet long, 5 pipes wide and 10 pipes high. In compartment No. 2 set up two box coils 5 feet long, 5 pipes wide and 10 pipes high. In compartment No. 3 set up three box coils 5 feet long, 5 pipes wide and 10 pipes high. *Heating and Ventilating Chamber.*

Support these coils on horizontal pieces of 1½-inch iron pipe let into holes bored in studding.

From side of compartment No. 1 and close to top take out the air-ducts L and K (see Fig. 2) for top or third floor.

From top of compartment No. 2 take out the ducts F, G, H, I, J, for second floor.

From top of compartment No. 3 take out the ducts A, B, C, D, E, for first floor.

Separate Cold-Air Ducts. Each compartment or chamber to have separate wooden cold-air duct, equal in area to the total area of warm fresh-air ducts supplied by it, connected to bottom of each air chamber (see Fig. 2).

Valves in Ducts. A nicely-fitting slide valve to be fitted in each cold-air duct to regulate supply. Cold-air duct to be taken from point shown on plan (Fig. 2). Each warm fresh-air duct to have a valve to regulate supply close to chamber.

Cold-Air Inlet. Build in wall where shown on plan (Fig. 2) a wood frame 40 x 14 inches, extend through wall, divide on inside of wall into three ducts and carry each duct to its respective chamber or compartment. No. 1 duct, 14 x 7 inches; No. 2 duct, 14 x 12 inches; No. 3 duct, 14 x 20 inches. Make a sash to fit frame flush with outside of wall and cover with wire cloth, ⅛-inch mesh.

Size Cold Air Ducts.

Warm-Air Ducts. The warm fresh-air ducts to be laid from heating chamber to registers in rooms, as shown on plan, Fig. 2, as direct as possible, avoiding acute angles.

Size of Ducts The ducts to second and third floors to be carried up in partition walls, using side registers in rooms.

The ducts to be of the following sizes :

	Horizontal pipes.				Square vertical pipe.
10 x 14 Register. Duct A.— To parlor...............	64 inches or 8-inch pipe (round)				......
10 x 14 Register. Duct B.— To library...............	49	"	7	"	"	......
10 x 14 Register. Duct C.— To dining-room........	49	"	7	"	"	......
12 x 15 Register. Duct D.— To hall, 1st floor..........	49	"	7	"	"	......
10 x 14 Register. Duct E.— Entrance north and toilet..	64	"	8	"	"	......
10 x 14 Register. Duct F.— N. E. Chamber, 2d floor....	36	"	6	"	"	9 x 3½
10 x 14 Register. Duct G.— S. E. Chamber, 2d floor....	36	"	6	"	"	8 x 3½
10 x 14 Register. Duct H.— S. W. Chamber, 2d floor....	36	"	6	"	"	8 x 3½
6 x 10 Register. Duct I.— Bathroom, 2d floor........	25	"	5	"	"	7 x 3½

	Horizontal pipes.	Square vertical pipe.
10 x 14 Register. Duct J.— N. W. Chamber, 2d floor.	40 inches or 6½-inch pipe (round)	11 x 3½
10 x 14 Register. Duct K.— Playroom, 3d floor.....	49 " 7 " "	10 x 3¼
10 x 14 Register. Duct L.— Billiard and chamber, 3d floor.	49 " 7 " "	14 x 3½

Ducts E, J and L to be laid as below described. Duct E, lay an 8-inch round pipe from top of heating chamber, as shown in Fig. 2, to a point near to and under register E on first floor, north entrance. From this pipe lay a 5-inch pipe to toilet-room, and connect with side register 6 x 10, where shown on Fig. 3. Reduce duct E at 5-inch connection, and continue 7-inch pipe to 10 x 4 register E in N. entrance. Fit in regulating valves on register side of connection. Duct J, lay a 6½-inch pipe in basement, and carry up in partition to N. W. chamber 11 x 3½, where shown in Fig. 4. Fit in double register box, Fig. 11, with one register 6 x 10 in linen-room, and one in N. W chamber 10 x 14. Duct L, lay a 7-inch round pipe in basement, and carry up in partition, where shown on plan, Fig. 2, 14 x 3½ pipe to second floor. There divide pipe, taking 7 x 3 pipe to partition running at right angles, as shown by dotted lines L L, on second floor, Fig. 4, and carry up to chamber register 6 x 10, third floor. Continue the other pipe up to billiard-room register, 8 x 3½ pipe 10 x 14 register

	Inlet.		Outlet.	
Parlor register A, 10 x 14...............	1	Register.	1	Register.
Library register B, 10 x 14..............	1		1	
Dining-room register C, 10 x 14........	1		1 =	6
Hall register D, 12 x 16...............	1		=	1
Bath and toilet, linen-room, 3d floor, 6 x 10..	4		4 =	8
Remaining registers, 10 x 14..........	6		6 =	12
Total...			27	

The chimney flues, where they are situated in rooms, to be used for foul-air outlet, setting in registers close to ceiling, of same size as fresh-air inlet.   Ventilating
Flues.

From bathroom, sewing-room and northwest chamber lay pipes from close to ceiling same size as fresh-air inlet, and connect to 8-inch diameter pipe in attic, and carry through roof near kitchen flue, as in Fig. 5. Cap with approved vent top, and fit casing round pipe similar to air pipe of expansion tank.

Attic Ventilation. Attic may be ventilated by means of registers fitted in "over window." Could be opened or closed by pendant cords.

Cellar Ventilation. A 6-inch flue should be carried down to vegetable cellar, and a 6-inch fresh-air inlet, with slide valve, built in wall at opposite end of cellar. Inlet may be 6-inch cast-iron pipe in wall, with hub on outside end. Over hub end fix a piece of wire cloth, and on inside fit a galvanized-iron slide valve.

Toilet-Room The foul-air outlet, should it not be provided for in plumbing work, may be carried up to third floor and taken through roof, with casing, &c., similar to 8-inch pipe.

Air Ducts and Register Boxes. Air ducts and register boxes to be made of IX tin, with tight joints, and put up in a careful and workmanlike manner.

ESTIMATE OF TINNER'S WORK.

Ventilation Registers.

North entrance: Parlor, library, dining-room, northeast chamber, southeast chamber, northwest chamber, southwest chamber, billiard-room, playroom, eighteen 10 x 14 registers	$36.00
Hall, one 12 x 15 register	2.50
Toilet, bath, linen-room, chamber, 3d floor, eight 6 x 10 registers	12.00
	$50.50
Tin for lining air chamber, IX, 20 x 28	6.75
12 IX tin air ducts, valves and register boxes	80.00
Outlet ducts, bath, toilet, linen-room and 3d floor chambers	20.00
Nails, $1; solder, $1.50	2.50
12 days' labor, tinner and helper, @ $6.50	78.00
Total	$187.25
Add—Galvanized-iron smoke-pipe to boiler	$10.00
Register, 7-inch round, chain, weights, &c	4.50
Labor fitting smoke-pipe and draft inlet	3.00
	17.50
Total	$255.25

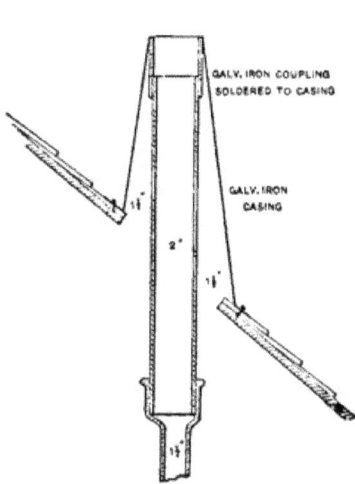

Fig. 8.—*Enlarged View of Air Pipe.*

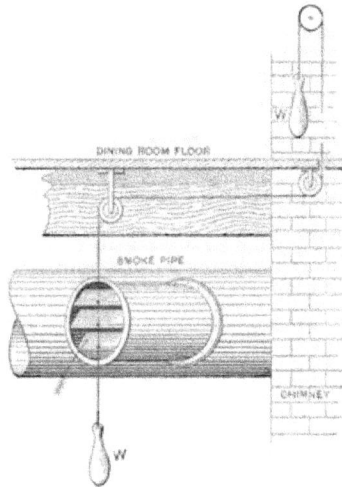

Fig. 9.—*Air Inlet Draft Check.*

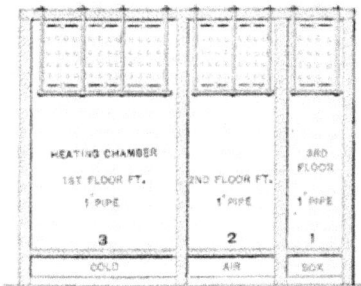

Fig. 10.—*Heating and Ventilating Chamber.*

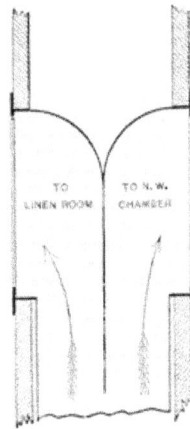

Fig. 11.—*Double Register Box.*

Heating Estimate.

1 Hot-water boiler	$245.00
665 feet of cast-iron radiator	221.15
1500 lineal fe t box coils of 1-inch iron pipe at 11 cents per foot	165.00
640 lineal feet wall coils of 1-inch iron pipe at 9 cents per foot	57.60
2 2½-inch Peet valves	15.00
9 2-inch Peet valves	22.50
19 1¼-inch angle valves	23.75
15 1-inch angle valves	15.00
2 1½-inch Peet valves	5.00
20 ⅜-inch draw-off valves at 40 cts.	8.00
32 ¼-inch air-cocks at 15 cents	4.80
1 1-inch check valve	1.00
20 ⅜-inch nipples	.60
9 2½-inch "	.56
18 2-inch "	2.16
6 1½-inch "	.54
28 1¼-inch "	1.96
50 1-inch "	2.50
10 ¼-inch "	.25
2 2½-inch R. and L. couplings	.60
10 2-inch R. and L. "	1.40
6 1½-inch R. and L. "	.60
30 1¼-inch R. and L. "	2.40
50 1-inch R. and L. "	3.00
100 1-inch elbows	6.00
58 1¼-inch "	4.64
10 1½-inch "	1.50
Carried forward	$812.51
Brought forward	$812.51
20 2-inch elbows	4.00
5 2½-inch "	2.25
20 ⅜-inch "	.70
10 ¼-inch "	.25
5 1-inch Peet valves	6.25
2 2½ x 2 x 2 Ys	1.50
12 2-inch reducing Ys	4.20
16 1½-inch reducing Ys at 25 cts.	4.00
10 1¼-inch reducing Ys	1.80
2 1-inch tees	.15
2 1¼-inch tees	.20
35 feet 2½-inch pipe	12.95
85 feet 2-inch pipe at 20 cents	17.00
150 feet 1½-inch pipe	15.00
450 feet 1¼-inch pipe	40.50
400 feet 1-inch pipe	24.00
75 feet ⅜-inch pipe	1.87
25 feet ¼-inch pipe	.50
Wrought-iron hangers	5.00
1 Sink or copper receiving pan for draw-off	3.00
1 Expansion tank complete	5.00
Wall stays for coils	8.00
Galvanized iron	1.00
Oil, red lead	2.00
Asbestos and hair felt	9.00
30 days' labor man and helper at, say, $6.50 per day	195.00
Total	$1,177.63

HOT-WATER CIRCULATION.*

BY JOHN HOPSON, JR.

For *The Metal Worker* competitions in house heating I desire to submit the plans hereto attached, viz.: First, second and attic floor plans showing locations proposed for **direct** radiators, their superficial area in square **feet**, hight and the position of flow **and return** risers and their connection with the radiators. Also on **first-floor** plan the position of registers for ventilation **hereafter to be proposed.**

In the cellar plan, the position of the boiler and its flue connection with house chimney, the flow and **return** mains to and from the risers above-mentioned, the location and connection with the boiler of the stack of indirect radiators and its outer air supply and warm air discharge, all forming a plant of hot-water heating apparatus to be open to the atmosphere at the expansion tank, and properly designated as a low temperature water heating apparatus.

The plans above stated form and are made a part of these plans and specifications.

SPECIFICATION.

For the attic floor, Fig. 4, I would place in the children's playroom 72 square feet of heating surface, placed by means of a pipe radiator on the outer walls as shown, using 180 lineal feet of 1¼-inch wrought-iron pipe, and formed into a radiator of six pipes in hight by means of branch tees, one placed at each extreme end and another branch tee formed to take the six pipes from adjoining right-angled faces, tapped for the purpose and made up as shown in the plan.

Attic Playroom Radiator.

The purpose of this corner branch tee is to take the flow from riser B, as shown, and distribute the water in either direction to the branch tees at the ends, which, being connected at the run on the bottom, one with the return riser T and the other with P, complete the circulation.

The flow connection should be 1-inch pipe and enter the room on the floor, being brought from the position of riser B on the floor and in the space formed by the roof shown in elevations. This pipe and all others similarly laid on this floor I would protect from loss of heat in same manner as that to be specified later for mains in the cellar.

* From *The Metal Worker*, April 20, 1889. Copyrighted, 1889, by David Williams.

An angle valve should be placed in this flow-pipe to form connection with it and the bottom of the angle branch tee, by which the heat can be regulated or shut off, if desired.

Billiard-Room Radiator.

I would place in the billiard-room a wrought-iron pipe radiator, as shown in Fig. 4, to contain 44 square feet heating surface, made up of 96 lineal feet 1¼-inch pipe of dimensions and with connection as shown therein.

Attic Chamber Radiator.

I would place in the chamber a wrought-iron pipe radiator of 26 square feet heating surface, made up of 56 lineal feet of 1¼-inch pipe of dimensions and with connections as shown,

Attic Hall Radiator.

and for the hall 17 square feet, made from 1¼-inch pipe and placed and connected as shown.

Second-Floor Radiators.

For the second floor I would place radiators of dimensions, capacity and connections as shown by plans of this floor, Fig. 3—viz.:

Heating surface.

Alcove... ....Cast-iron radiator, low pattern,	15 sq. ft.
Parlor chamber } .... " " " "	45 "
....	33 "
Library chamber........ " " " "	66 "
Dining-room chamber } . " " " "	30 "
. " " " "	30 "
Sewing-room.......... " " , " "	24 "
Kitchen chamber.....Cast-iron radiator, medium hight,	35 "
Staircase hall.................Wrought-iron pipe coil,	31 "
Bath-room.................... " " "	9 "

First-Floor Radiators.

For the first floor I would place radiators of dimensions, capacity and connections as shown on plans of this floor, Fig. 2—viz.:

Heating surface.

Front hall.............Cast-iron radiator, medium hight,	35 sq. ft.
Parlor { .... Cast-iron radiator, low pattern,	24 "
{ .... " " " "	24 "
{ .... " " " "	39 "
{ .... " " " "	39 "
Library { .... " " " "	66 "
{ .... " " " "	30 "
Dining-room } .... " " " "	33 "
{ .... " " " "	51 "
Butler's pantry......... " " " "	15 "

Hot Closet.

and attach thereto, on top, a hot closet of dimensions 14 inches deep, 18 long and 14 high, for the purpose of warming plates, &c.

Rear Hall and Toilet Radiators.

For want of space in the rear hall, to make other disposition of heating surface, I will take from flow riser F, Fig. 2, at the ceiling in the toilet-room, a branch to run, as shown, to a branch tee placed horizontally, from which four 1¼-inch

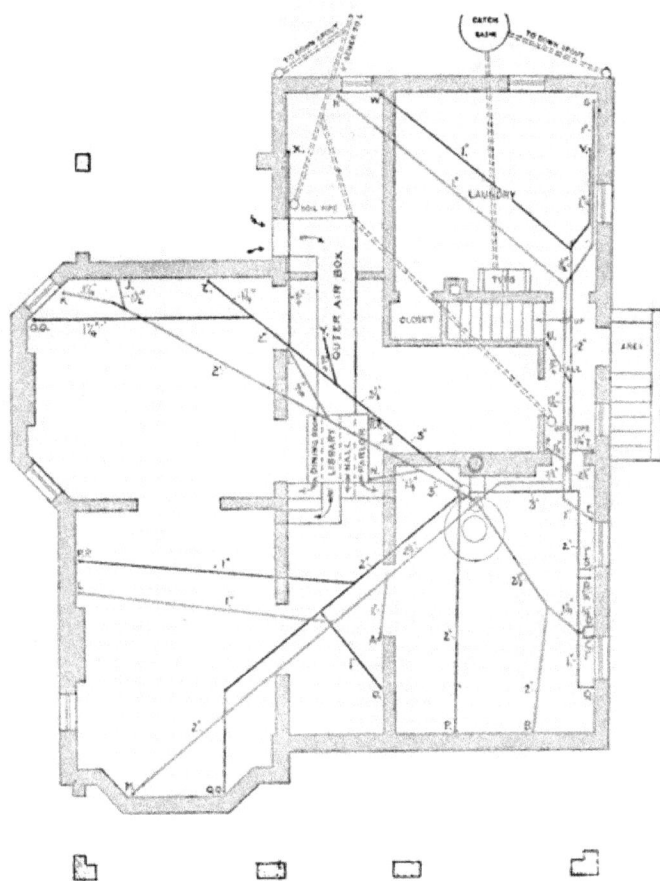

Fig. 1.—Cellar Plan.—Scale, 1-12 Inch to the Foot.

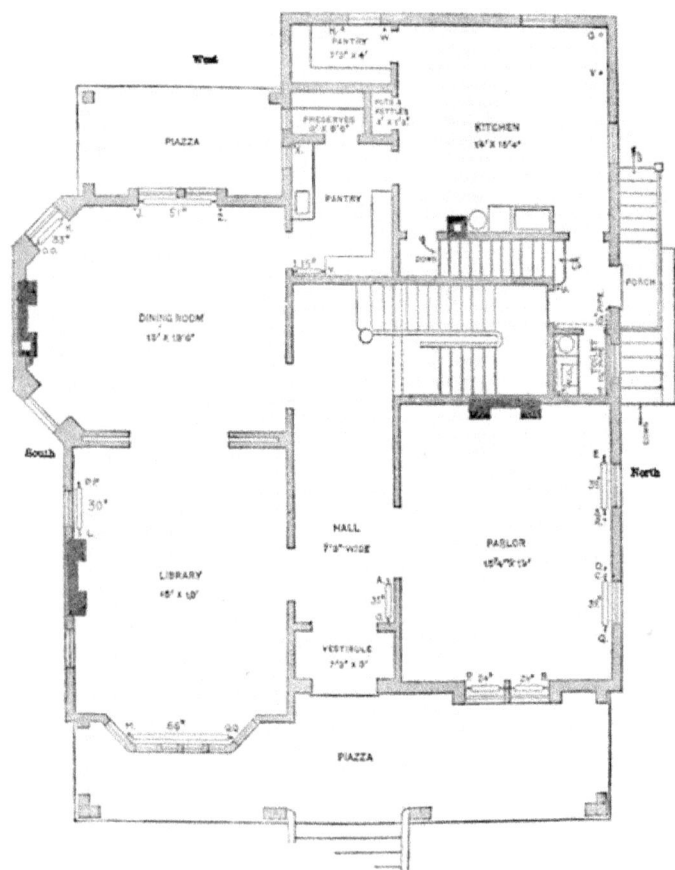

Fig. 2.—First-Floor Plan.—Scale, 1-12 Inch to the Foot.

pipes are taken to run down the wall, in position as shown, to a point where, with suitable spaces between them, they may elbow through the partition into toilet-room and run through it under the window to a branch tee placed vertically in the northeast corner, to take the water in this circulation, and by return pipe led from its run at the bottom discharge into the return main.

A gate-valve placed under the return branch tee will regulate this circulation, and the flow-pipe may incline slightly downward from the point of leaving the riser, and this radiator will then find vent through the flow.

The heating surface hereby had in the back hall is 13 square feet, and in the toilet-room is 10 square feet.

I would place in the cellar, suspended from the floor beams overhead, 13 Gold pin indirect radiator sections, supported by wrought-iron hangers. These sections to be of 9 square feet heating surface each, or a total of 117 square feet, and connected with the circulation by one 1¼-inch pipe and fittings, as shown in the cellar plan, Fig. 1. *Ventilation*

This stack and outer-air and warm-air ducts to be inclosed by one 1¼-inch plank and firmly bolted together and lined with tin, the opening to outer air to be covered with strong copper netting, and any opening about the stack through which air could pass without being warmed to be stopped with strips of tin. The ducts leading from out-of-doors to be 4 feet wide and 1 foot high in clear and tin partitions being run through the hot closet, as indicated in dotted lines, partitioning its contents proportionally to the cubic space in each room to be supplied with ventilation. *Stack.* *Cold-Air Ducts.*

I would lead from it to the several rooms ducts and registers as follows: *Ventilating Registers.*

Hall, to supply first and second floors.....Duct 200 sq. in., register 14 x 22
Parlor.......................... ...   " 170   "   "   14 x 22
Library................... ....   " 110   "   "   12 x 15
Dining-room............................   " 110   "   "   12 x 15

These registers may go into the walls if owner prefers, and will be finished in copper and brass plate or at the extra expense of the owner if higher quality finish is desired.

The sections composing this indirect stack should have connections at either end one with another, and the flow entering, if at the corner nearest the boiler, as shown, the return should be at the other end of the same side, and the radiator to be set to incline slightly downward from the flow to the return and the cold-air space under it and warm-air space

Total Heat-
ing Surface. above incline conformably therewith, the total heating surface herein being:

In the attic ... . 159 square feet.
Second floor.............. ' 318   "   "
First floor.................................... 373   "   "
Indirect stack.... 117   "   "
And in risers and connections not protected by
      covering from loss of heat amounting to....... 115   "   "

Or a total of 1088 square feet for the entire house. The
selection of a boiler of suitable capacity follows.

Boiler. In a good cast-iron boiler, constructed to present effective
heating surface to the combustion, and the parts so disposed
and connected as to offer free and quick circulation, 1 square
foot of boiler will carry, in my experience, 10 square feet of
radiators when the capacity of the radiators themselves have
been determined by careful calculation and are sufficient to
warm the house.

Boiler
Surface. I will then select a boiler of 110 square feet heating surface
answering the conditions above stated. It should have an
ash-pit of not less than 12 inches in hight—14 would be
better—and of area equal to the full area of the combustion
chamber.

Grate. The grate is preferably round and of 3 square feet area,
exclusive of the area occupied by fresh coal discharged from
the self-feeding coal magazine, if one is supplied. I would
provide for such a boiler a grate to be shaken by a lever outside and containing 2 square inches air space to 1 square inch
of iron.

Water
Sections. The sections comprising the boiler and their connections
with each other should be such that the water may rise
freely when heated toward the point of exit, and the means of
making such connections should be by wrought-iron threaded
nipples as the most durable and best joint.

Planed faces, to contain gum or asbestos packing or rust
joints, are not to be approved, and herein constitutes an item
of expense which I consider justified by the value in cost of
the connections I mention over those I do not approve.

The boiler sections should be so disposed as to permit the
freedom of expansion and contraction without strain to any
other part, and these parts should not be tightly bolted
together.

Combustion
Chamber
Flues. The combustion chamber flues, for the conduct and escape
of gases, should be so arranged and disposed and built of
suitable capacity as to make full use of the heat generated
and the gases not be let to escape from the boiler or its surroundings, or contained flues at a temperature much, if any,
above that of the water warmed.

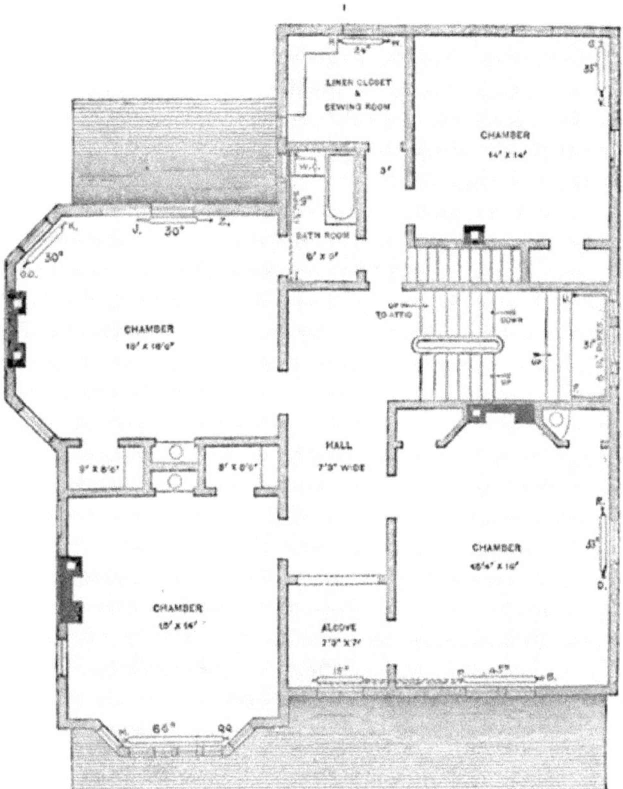

Fig. 3.—Second-Floor Plan.—Scale, 1-12 Inch to the Foot.

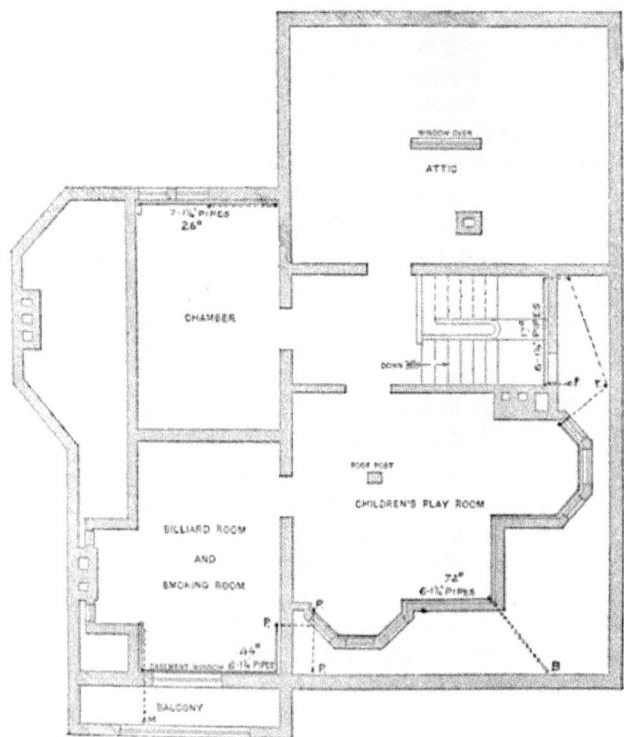

Fig. 4.—Attic Plan.—Scale, 1-12 Inch to the Foot.

For the boiler specified, an 8-inch flue, leading from the **Boiler Flue.** boiler to the chimney, would be suitable and preferably constructed of copper. The flue in chimney should be an 8-inch tile leading from the cellar to the top of chimney, inclosed in the brickwork of which the chimney is constructed and provided with a cleaning-hole at the cellar floor. This flue should have no other connection or opening between the boiler and the top.

This boiler may be either set in brick or of portable pat- **Boiler Setting.** tern, provided that in the portable pattern the casings inclosing it shall be double, the space between them not less than 1 inch in thickness and filled with asbestos, mineral wool or other good and durable non-conducting material.

No part of the boiler, whether brick-set or portable, should be left exposed to the air of the cellar, but should be well protected from loss of heat. Access to all interior and exterior flues should be readily had for the purpose of cleaning.

As to a magazine feed, admitting the theoretical waste of **Magazine Feed.** coal, there is much compensation from a steady feed, and this being a private house, I know of no construction whereby a steady fire and its maintenance for length of time, particularly at night, can be so uniformly attained. The chimney being suitably provided for draft, there will be no escape of gas from it into the cellar.

The boiler draft may be contained in the ash-pit door, and **Boiler Draft.** whether the feed magazine is used or not, a surface fire-door of suitable size should be provided.

A boiler as described above would be of about 42 inches **Size of Boiler** diameter if portable, or about 45 inches square if brick-set, and about 52 inches high.

Taking as central a position as can be had on top of the **The Cellar** boiler as a place of exit, I desire all the water to supply the **Mains and Risers.** several radiators to diverge from this central point placed at a level above the boiler suitable for the mains.

A fitting of proper size, with outlets at the points from which it is desired to run the mains and tapped to take them, can be had. It is also important that all mains leave from the same level, and that no main have a plainer course to lead the water through than any other.

The vertical pipe leading from the boiler to supply the mains should be of 4½-inch size; also the return to boiler from the point near it where returns meet. Thus the flow and return mains are overhead, and all flows are to be run at an incline upward from and all returns at an incline downward to the boiler.

The cellar mains (also any risers) not exposed to view **Covering for** above the first floor should be covered, to prevent loss of **Mains and Risers.**

heat, by asbestos paper first wrapped on, then ¾ hair felt, and then wound with stout canvas well sewn.

Location of Pipes.   On the cellar plans and throughout the other house plans I have designated the positions where I propose to run the several pipes, marked their sizes and shown the style of fittings to be used in making connections.

Size of Risers.   Beginning in the attic, it is first determined the area of the connections to the several radiators. This size of pipe so determined is then run as a riser to its point of connection with any radiators near it on the floor below, when the riser is enlarged in size to equal the sum of the areas of all the connections served, and so on.

Fittings.   In addition to this provision of area, care has been taken to use such fittings as will prevent an unequal service in preference of one radiator over another.

Supports for Cellar Mains.   The mains in the cellar to be supplied with ¾-inch hooks screwed into the beams and slung in No. 10 iron chain.

Floor and Ceiling Bushings.   The risers at the point of entering floors and ceilings to be provided with floor and ceiling bushings to cover the cuttings made.

In addition to the plans heretofore mentioned, I append charts, Figs. 5 and 6, showing the risers and their connections to further illustrate this system.

In order to comply with your expressed wish to be able to shut off the heat from each room, I would supply each radiator with a nickel-plated wood wheel angle valve placed on

Valves.   the flow ; also, would place in the top of each radiator a nickel-plated air-cock, to open with key, for the escape of air.

Also on the indirect radiator a gate-valve, to close this circulation when desired.

Expansion Tank.   There will be contained in the boiler and radiators and pipes in this plant about 180 gallons of water.

Capacity of Tank.   Water expands from 39° to 212° 1 gallon in every 25, approximately, and to contain this full limit of expansion the tank should contain 7⁷⁄₆ gallons in sight on the glass gauge, or of 8 gallons capacity.

Material   I would select this tank of cast iron and place it in the attic hall some 4 or 5 feet above the top of any radiator on that floor.

Gauge.   A glass gauge should show the hight of water in the tank. The top should be covered and a 1-inch pipe led from it as an overflow-pipe to the sink or waste-pipe you speak of in your inquiry.

Overflow.   This overflow should contain an air-cock at its highest point to break the syphon in case of overflow.

Expansion-Pipe.   The expansion-pipe, 1 inch in size, should lead from the boiler to enter expansion tank at the bottom.

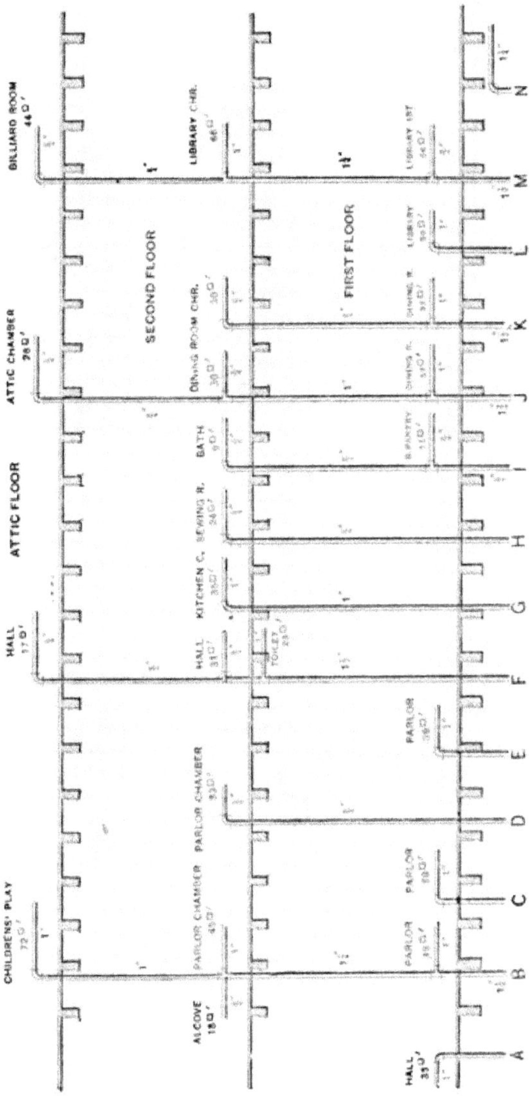

Fig. 5.—Chart of Flow Risers, Showing Decrease in Size.

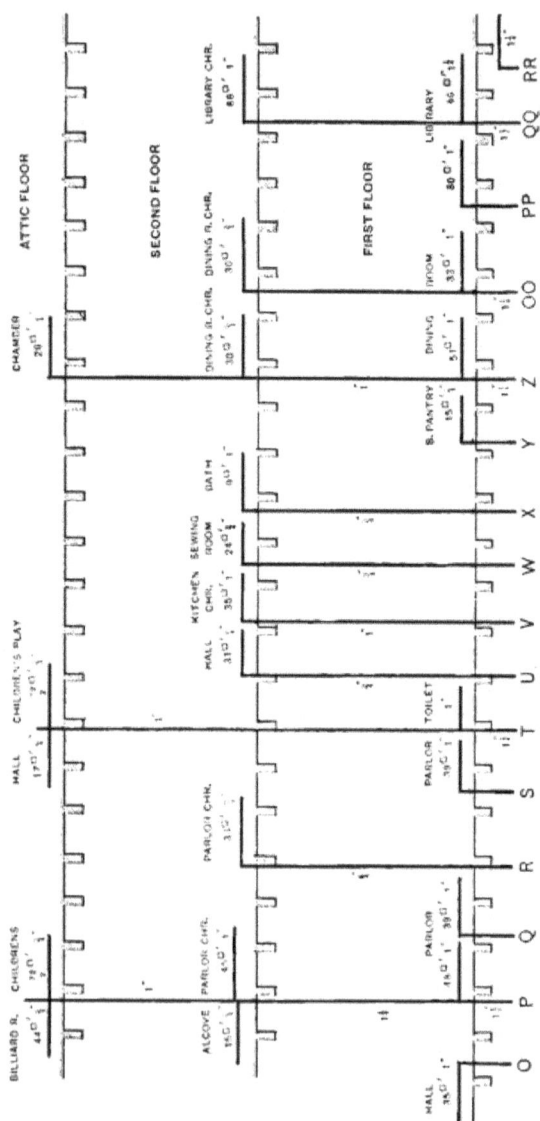

Fig. 6.—Chart of Return Risers, Showing Decrease in Size.

You state the water-supply for the house is in the attic, but I do not know whether it stands above the position I selected for the expansion tank or not. If above, I would connect this supply with a ball-cock contained in the said expansion tank to fill the plant automatically and shut off the supply at a point below the 8-gallon capacity mentioned. Water-Supply.

This supply should have a valve to shut off between the tank and ball-cock.

In case the attic tank will not permit this arrangement, or the city pressure will not deliver water to the aforesaid ball-cock, then I would connect the water tank with the expansion-pipe by means of a $\frac{3}{4}$-inch pipe, containing a valve to close when the water had flowed into the apparatus to its level, and by means of another pipe connection and a small force pump fill the expansion tank to view in the glass gauge.

I would place in the flow at point of exit from the boiler, also at the point where the return mains deliver to the boiler, also in some flow riser in a convenient place in the house, as in the rear hall or the dining-room, thermometers which accurately and legibly indicate the temperature of the water in circulation. Ther-mometers.

Nothing exists but that it can be made to do better by intelligent rather than by ignorant use. The use of these thermometers is education toward the best use of the plant, and this is precisely the arrangement I would like in my own house.

If the position of this last-named thermometer is near the handle mentioned, under heading of automatic damper regulator, the convenience of both arrangements will be apparent.

The use for an automatic damper regulator in a hot-water heating apparatus is not only to regulate the drafts suitably to the heating required, but also to prevent the water from boiling. Damper regulators working from the expansion of a metal have often proved unreliable from the fact that no metal after being expanded by heat contracts to its original length. Automatic Damper Regulator.

Damper regulators for this purpose, which operate the drafts from the expansion of a liquid more susceptible to expansion and contraction than water, and when immersed in it in a copper vessel, will accomplish the purpose desired, and such a one I can recommend by my experience, and would specify it to be placed in the flow at the point of exit from the boiler and connected with the draft and check dampers by use of a suitable lever, chains, &c. This lever may also be connected by chains with a handle to be fixed at some convenient place on the first or second floors, whereby the point at which the automatic regulator opens or closes the drafts may be changed, or its operation to open the drafts, but not to close them, may be superseded at the pleasure of the occupant of the house.

Bronzing.   All pipes in view above the first floor and all radiators, expansion tank, &c., should be covered with a good coat of copper, orange or gold bronze, as the owner may select.

Blow-Off.   A blow-off, consisting of 1-inch pipe, with a 1-inch steam cock, to discharge into the waste-pipe you mentioned, should lead from the bottom of the boiler.

ADVANTAGES AND GENERAL FEATURES.

Having set forth and specified the work I would propose for warming the house, I will now comply with the condition you make that the reasons why shall be stated.

One reason why I prefer to submit a plan for warming the house by hot water instead of by hot air or by steam is that, having had experience in the use of all these in different houses I have lived in, I consider hot water preferable to any of the others.

Public opinion or common talk has always been recognized as a factor in determining the truth, and the progress which hot-water heating has made among users in the past few years is an argument that the growth of the art has evolved its merits.

It is open to the atmosphere and absolutely safe; it is also noiseless. Technically I view it as the best, because the source of warmth to a room or house either by the direct or indirect method is at a lower temperature than that of hot air or steam. The heat from it by either method is often designated as being soft and mild. Its low temperature changes the quality of the air as to humidity less than either of the others. This reason of a lower temperature, together with the fact that water is a steadier and more reliable vehicle for the conveyance of heat, forms the chief reason why it is more economical than any other.

It therefore accounts at the radiator and in the place where it is intended to be used for more of the units of heat generated by the fire, and is the most economical in the use of the fuel consumed. It is my belief that while the hot-water plant will cost the most of the three, even if all are as well erected of their kind as they may be, the hot-water plant will prove the most economical in the end, maintenance and repairs being taken into account.

The reason why I allot a particular quantity of heating surface to each room forms also the reason why I designate a particular position taken for the radiators, and to this I desire to call your attention. The house if heated to 65° and 70° in zero weather, as desired, in all its parts may be viewed as a radiator losing heat to the outer air. To supply this loss it is pertinent to observe at what particular points the loss takes place and at what rate. For the purpose of illustration one room will suffice.

Take the parlor on the first floor, Fig. 2. This room contains 283 square feet of outside wall, 77 square feet of glass and 4095 cubic feet air space. The hall walls surrounding it, being warmed to 70°, are *nil* as to loss of heat. The entire loss takes place, then, through the walls and glass and by the change of air entering the room through cracks and crevices, and by reason of the exit of the warmed and expanded air contained in it. The loss of heat per square foot of wall constructed as you describe it is, in units, 0.223 per square foot per hour for a difference of 1° between the inner and outer walls. You wish to maintain 70° in this room, and in order to do so must provide for the loss of 70 times the loss of 1° per square foot per hour, or a total of 4418 heat units. For the glass the loss of heat in units is 0.543 per hour per square foot for a difference of 1°, and this by 70 into the total square feet of glass in the room requires a total supply of 2927 heat units per hour for this loss. The incoming air affords a field for the use of judgment and experience. The house is not air-tight.

It is not possible to say from your description, and no one can determine accurately before the house is built, or after it, the precise quantity of air that will enter from zero out-doors to be warmed to 70° inside, under the varying conditions of wind pressure. All I can say about it is that from such judgment and experience with this method of estimating surface required in a house, and the satisfactory results I had obtained elsewhere, your description of the construction, location and exposure of the premises leads me to assume with some degree of confidence that if the heating surface be provided in this respect with a change of air equal to twice the cubic contents of the room it will not only be well warmed, but also well proportioned to the other rooms of the house, a matter of much consequence. A cubic foot of dry air at zero weighs 0.0924 pound, and requires 238 units of heat to warm it 1°; we wish to warm 8190 cubic feet (70°), and for that purpose must expend 12,551 units of heat.

Now, you wish to entertain plans for an apparatus which will warm the house to 70° in zero weather without forcing.

In this low-temperature apparatus we can maintain an average temperature of 180° to 200°, but the latter figure would require forcing and careful attendance. The most satisfactory work I have planned has been to warm from a circulation of 160° as an average temperature of the heated body, and I will take that for this plan.

The iron radiators commonly used will emit, when standing in a room, 156 heat units per hour—of this quantity 126 by contact of the air with them and 30 by radiation. The total loss of heat being found to be 19,896 heat units per hour, 128 square feet radiators is the necessary quantity, and hence I apportion this amount to this room. ·

It now becomes a question of much moment as to where this radiator shall be placed. Of the total demand for the supply of heat, one-seventh of it is at the window—in other rooms this proportion is larger —one-quarter, nearly, is through the outside walls, and the remainder to warm the incoming air, but this incoming air enters at the windows for the most part. An anemometer held at the window-sill in such a position as to take the down current will show at once that the entering air and down current of air cooled by the glass require that the work of heating shall be done under the windows, and I have placed the radiators in that position for this reason.

If the heating is done by radiator placed on the inner walls of the house this current of cold air must traverse the floor to it to be heated, and it is not enough at this day of the world to produce a given effect on a thermometer placed in the center of the room. You wish to make use of all of the rooms and the windows, and on a stormy day, as well as any other, you ought to be at liberty to occupy the library without feeling a draft of cold air on your feet. I can suggest no other suitable disposition of the heating surface except this, and in the light of the object sought I believe it is the best. I am asking you in this item to justify a considerable expense over the cost of placing one radiator near the boiler.

The work of setting these four radiators and their connections is almost four times that of connecting one radiator.

The pattern which I mention as a low pattern is one commonly made varying in hight from 16 to 20 inches. These are the most expensive form of cast-iron radiating surface in the market, but the justification of this expense lies in doing the work well. It may strike you as novel to place four radiators in a room of the size of the parlor. The number follows on the decision as to whether the need for such a position has been demonstrated. You may further object to them as an obstruction rendering the window and blinds inaccessible. My experience of the results in houses containing radiators so placed is that the position is acceptable rather than otherwise. Whatever changes of furniture you propose you would not place any of it under the windows, and a direct radiator elsewhere might be such an obstruction. These radiators are about 9 inches wide, and the space under them is open to receive the foot when standing near the window, and hence is much less of an obstruction than might be supposed. Again, you have not asked me to arrange the heating of the house conformably to any furniture spaces, but to present an ideal plan of heating. This I venture to do as shown.

These facts, as stated, apply to every room in the house. Radiator positions are chosen to apply the source of heat to the loss as nearly as the construction of the house and arrangement of the rooms will admit. In the library the estimated change of air is taken at one and one-half times per hour, and in the dining-room at one and one-quarter—owing to points of compass toward which they face and in view of prevailing cold winds. A house so proportioned as to heating surface should be much more uniformly warmed than one apportioned from cubic contents only. I have provided a means of apportioning the surface directly for each square foot of outside wall, glass, and per cubic foot of an assumed change of air.

It will be seen I propose nothing which does not demand the same attention from direct heating by steam.

In the butler's pantry, in order to take the hot-closet radiator, I will ask that the door be moved about 8 inches toward the sink. The risers in the kitchen pantry may be covered and closed if it is desired to keep the pantry cool. I consider it important to place the radiator on the stairway landing, from the cause above stated, and to prevent the down draft of cold air particularly noticeable about well-holes in halls on a windy day, and to do so (for want of space) use a pipe radiator.

The pipe radiators are somewhat less expensive than the low-pattern cast-iron radiators, and therefore I use them on the third floor and particularly in the children's room, because the exposure of the outer wall is more there than in any other room in the house, and I wish to dispose the surface along the wall conformably to the animus of the plan.

It will be noted that care has been taken to specify that this apparatus is one of a low temperature as distinguished from high temperature or one under pressure. This uses no safety valve; that one does. This calls for an outlay necessary to warm the house to 70.65°, heated, &c., from a temperature of 160° in the heated body.

High temperature has no limit to the pressure or temperature of the heated body, and in weather requiring the use of its capacity is open to the same objections that are to be named for steam heating, as well as others.

The reason why I have introduced the given quantity of ventilation is that, although on extreme cold days while the air of the whole house is changing every hour to an amount quite in excess of the demand made by any reasonable number of occupants, there will be many days when the difference between the inner and outer temperatures will be small, with almost no movement in the outer air.

This ventilating stack may be then used as a very desirable addition to an ideal system. In it I provide for the artificial ventilation of the first and second floor halls, the parlor, library and dining-room to the extent of a change of air once an hour, or a total of 14,855 cubic feet per hour.

In order to cover all possible range of the need set forth I assume that this quantity of air at 32° may be required to be warmed to 82°, and there are required to warm it the expenditure of 14,261 heat units. Estimating from a temperature of 160° in the heated body, emitting 126 heat units per square foot per hour when heating by contact only, I find the indirect stack should contain 113 square feet heating surface.

In order that the velocity of the incoming air shall not much exceed 1 foot per second, and to provide against friction to retard the entrance, I would make the air space in outer air duct 4 square feet in area, and determine the areas of the ducts leading to the several rooms to deliver air proportionally to the contents thereof.

A damper to regulate or shut off the flow of air through the cold-air duct would also be provided, to be operated from the hall on the first floor.

I do not attempt to show in the charts, Figs. 5 and 6, more than the reduction in area of the risers due to the service they are to perform. It is not good practice to use smaller than ¾-inch pipe, hence the service of some of the smaller radiators on second and third floors seems out of proportion.

In all flow-risers the connections should be taken off from a tee, or where practicable from a Y, in order that no radiators shall be preferred in service by reason of the direction of the current.

I note that you state some of the windows have double glass. I infer you mean by this two panes of glass separated by an air space, but the protection in this instance is due to the thickness of the air space, and until I know what that space is I cannot say what reduction in loss of heat will follow.

I cite for authority on the conducting and radiating power of materials the findings of M. Peclet in "Manual of Heating and Ventilation," by F. Schumann, C. E.

I select 1¼-inch pipe for the pipe radiators, because it is my experience that its cost per square foot of heating surface and its effectiveness to deliver heat and at the same time maintain a fair average temperature render this size the most desirable.

All the pipe radiators should be made up with branch tees to form the ends, that the circulation may be in one direction only, except in the bathroom the radiator may be built with return bends.

ESTIMATE.

I estimate for the cost of this work as follows:

Boiler, of 110 square feet heating surface, either portable or brick-set.............	$232.00	Fittings, branch tees and bushings, pipe hooks and chains...................	$64.50
564 feet low-pattern radiators.	250.64	Non-conducting covering....	45.00
70 feet medium-pattern radiators...................	26.18	Bronzing...................	20.00
13 indirect radiator sections..	29.25	Registers................	25.11
8-gallon expansion tank, cast iron, with glass gauge....	12.40	Tinsmith, labor and material.	35.00
3 thermometers...........	13.20	Carpenters, labor and material...............	15.00
Automatic damper regulator.	27.50	Steam fitters, labor 2 men 20 days, at $7.50...........	150.00
Pipe used in job..........	92.32	Freight and cartage.......	25.00
Valves and air-cocks.......	36.00	Total.................	$1099.10

HOT-WATER CIRCULATION.*

BY G. C. BLACKMORE.

ADVANTAGES AND GENERAL FEATURES.

In submitting these plans and specifications I have a few things to say in reference to the completion of the work, and also my opinion of the way in which the work is to be done. In the first place, you do not allow us to mention any one make of heater, but I would prefer placing a cast-iron sectional heater with the amount of surface and grate-surface mentioned in specification and with a free circulation and as little friction to the water as possible.

My rule for ascertaining the size of heater is to find the square feet of radiating-surface required to heat the building to the desired temperature; then for every 15 square feet of radiation take 1 square foot for boiler, and for every 20 square feet of boiler take 1 square foot for grate. The rule in general use is 1 square foot of boiler-surface for 10 square feet of radiation, but with direct fire-surface I figure 1 to 15. In figuring up the radiation for a building, I calculate from the exposures, as you will see by following my figures. Dining-room, 2810 cubic feet, first floor, west and south exposure, four windows, 35 square feet surface to 1000 cubic feet air space. Library, first floor, 2335 cubic feet, south and east exposure, five windows, 33 square feet of surface to 1000 cubic feet of air-space. Main hall, first, second and third floors, 6442 cubic feet, 15 square feet direct to 1000 cubic feet, and 30 square feet of indirect radiation to 1000 cubic feet. Parlor, 3055 cubic feet, first floor, north and east exposure, 17 square feet surface direct to 1000 cubic feet, and 34 square feet to 1000 of indirect surface. In the pantry I propose making a coil of 1-inch pipe to be used as a plate-warmer and also for the purpose of heating same. Parlor chamber, 2327 cubic feet, north and east exposure, using direct-indirect, will require 40 square feet of surface to 1000 cubic feet. Dining-room chamber, 2641 cubic feet, south and west exposure, direct-indirect, 37 square feet to 1000 cubic feet.

Library chamber, 1995 cubic feet, east and south exposure, four windows, using direct-indirect, 35 square feet to 1000 cubic feet. Alcove front, 484 cubic feet, 30 square feet to 1000 cubic feet. Bath-room, 427 cubic feet, to be heated to 75°, requires 40 square feet to 1000 cubic feet. Sewing-room, 532 cubic feet, south and west exposure, 34 square feet to 1000 cubic feet. Kitchen chamber, 1862 cubic feet, northwest corner, direct heat, 35 square feet to 1000 cubic feet. Attic chamber, 1350 cubic feet, third floor, 28 square feet to

---

* From *The Metal Worker*, July 13, 1889. Copyrighted, 1889, by David Williams.

1000 cubic feet. Childrens' play-room, 2088 cubic feet, third floor, to be heated to 70°, north and east exposure, 35 square feet surface to 1000 cubic feet. Smoking-room, 1539 cubic feet, southeast exposure, heat to 70°, direct-indirect, 35 square feet surface to 1000 cubic feet. In rear hall I have left 1¼-inch riser exposed for warming same; also in back pantry for the same purpose, thus doing away with the necessity of running up risers in outer wall, while using their surfaces as radiation.

My object in placing indirect radiation for first floor hall and parlor is for the purpose of introducing a large volume of outer air, heated to a higher temperature than the air in rooms, for ventilating purposes. You will notice that I have also placed direct radiators in parlor and hall to heat same in conjunction with indirects; this I have found, in my experience, to give the best satisfaction. While a room or building can be heated entirely by hot-water indirect radiators, the same results can be obtained from a system as described above with a smaller consumption of fuel, while the ventilation would be as efficient as if entirely heated by indirect radiation.

In running mains in cellar I carry flow and return pipes side by side and take my connections for risers or radiators off top of same with nipple and elbow, reducing the size of main at each branch, thus securing an even and ample circulation to all radiators while the nipple and elbow allow ample provision for expansion and contraction. You will notice that I carry a separate main for first-floor radiators having no connection with risers leading to upper floors. This I find gives a more equal distribution of the heat, and allows the radiators on the first floor to heat as quickly as the upper floors, and effectually prevents upper-floor radiators robbing lower ones. In running second and third floor risers, when branching, I take the branch of the T for the third floor, and the end of the T for second-floor radiator. Where pipes pass through brick walls I use cast-iron sleeves larger than pipes and masoned in for the purpose of allowing free expansion and preventing damage to walls.

It is only necessary to place a single valve on flow-pipe of radiators, for the purpose of controlling circulation. I use nickel-plated elbows on opposite end of radiator to correspond with valve. My object for using keyed air-valves is to prevent any person operating them without a key. My reason for using slabs under radiators is that the carpets can easily be fitted to them and do not have to be cut to pieces to fit around the legs. Except for giving a nice appearance to the radiator a walnut slab will answer every purpose, but marble is generally used on account of looks.

The self-closing cock and funnel mentioned in connection with expansion tank is for filling by hand should the water-supply be shut off. The self-closing cock prevents water from boiling out in case person filling should forget to close an ordinary cock. The object of the

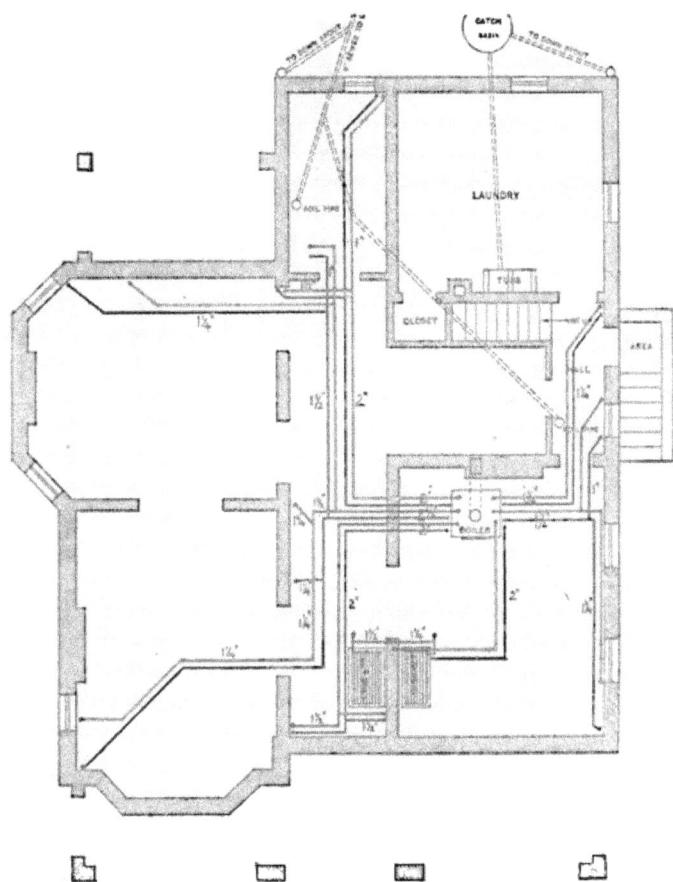

Fig. 1.—Cellar Plan.—Scale, 1-12 Inch to the Foot.

West

PANTRY
7'0" x 4'

RISER

PRESERVES
6' x 6'0'

POTS &
KETTLES
6' x 7'8"

KITCHEN
15' x 15'4"

PIAZZA

60' H 1" PIPE

PANTRY

1½"
RISER

101°

DOWN

1½"
Riser

PORCH

DINING ROOM
15' x 18'0"

VENT. REG.

VENT
REG.

South

100°

North

HALL
7'0" WIDE

PARLOR
18'4" x 19'

VENT. REG.

LIBRARY
18' x 18'

14'10½"

6' x 6"

VESTIBULE
7'3" x 9'

1½"
RISER

1½"
RISER

50°

85°

PIAZZA

East

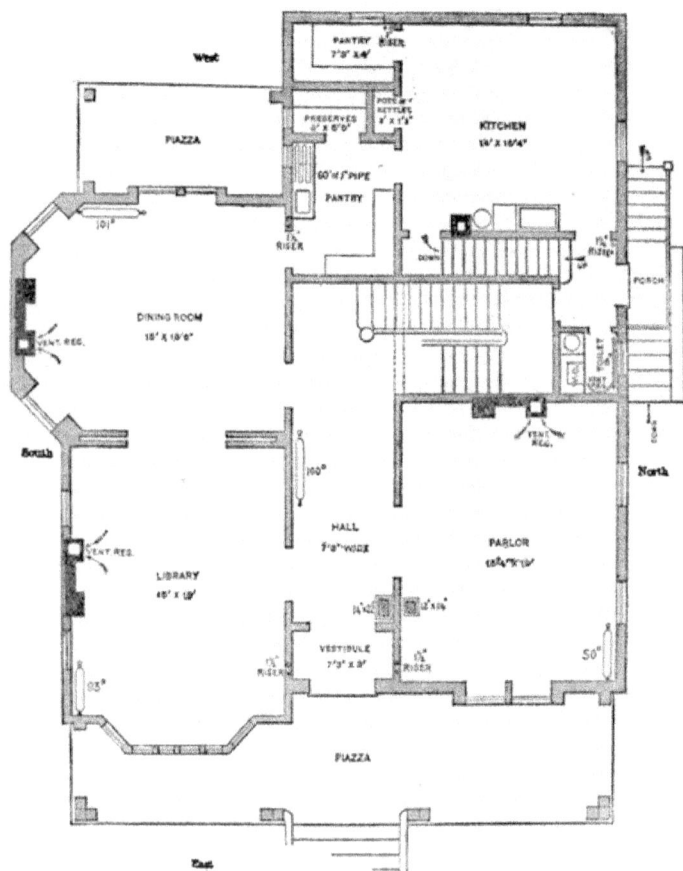

Fig. 2.—First-Floor Plan.—Scale, 1-12 Inch to the Foot.

check-valve between tank and water-supply cock is to prevent water already in tank from flowing back into supply-pipe should water be shut off.

In proportioning size of expansion-tank necessary I ascertain the amount of water in the whole apparatus and arrange tank of ample capacity to hold one-twentieth of the amount of water in the whole system.

In ascertaining the size of ventilating-flues I figure about 30 square inches of flue to every 1000 cubic feet of air-space, but varying according to hight of ceiling, putting in less where the ceiling is higher and more where the ceiling is lower

SPECIFICATION.

Place in cellar where marked on plan, Fig. 1, a hot-water **Heater.** heater containing 80 square feet of boiler surface and 4 square feet of grate surface with 9-inch smoke connection to chimney-flue (to be provided by owner) where marked on plan, and to have an internal area of 96 inches.

Heater to be connected to flue by 9-inch galvanized-iron **Smoke-Pipe.** pipe, 24 wire gauge, with round elbow and damper and with cold-air-inlet attachment.

Radiators to be placed where marked on plans, to be of **Radiators.** size mentioned below, as follows: Parlor, 100 square feet of indirect and 50 square feet of direct radiation. Main hall, 180 square feet of indirect and 100 square feet of direct radiation. Library, 93 square feet direct radiation. Dining-room, 101 square feet direct radiation. Kitchen chamber, Fig. 3, 65 square feet direct. Sewing-room, 17 square feet direct radiation. Bath-room, 17 square feet direct radiation. Alcove, 15 square feet direct radiation. Dining-room chamber, 98 square feet direct-indirect radiation. Parlor chamber, 92 square feet direct-indirect radiation. Library chamber, 70 square feet direct-indirect radiation. Children's play-room, Fig. 4, 73 square feet direct. Smoking-room, 53 square feet direct-indirect. Attic chamber, 38 square feet direct.

Place in cellar below main hall one stack of indirect, con- **Indirect** taining 180 square feet of plain surface, and a stack below **Radiators.** parlor containing 100 square feet plain surface.

Place in parlor chamber one direct-indirect radiator con- **Direct-** taining 92 square feet plain surface, one in library chamber **Indirect** **Radiators.** containing 70 square feet plain surface. Dining-room chamber, 98 square feet plain surface. Smoking-room, 53 square feet plain surface; these to be fitted with cold-air pipes, dampers, &c.

Run from ceiling of front piazza between second-floor **Cold-Air** joists to beneath direct-indirect radiators a galvanized-iron **Pipes.** pipe with an area of 23 square inches for parlor chamber and one with an area of 17 square inches to library chamber, also one from back piazza ceiling for dining-room chamber, with an area of 24 square inches, and one from front balcony to smoking-room, with an area of 13 square inches. Run galvanized-iron pipes from outside wall to below indirect stacks with an area of 135 square inches for main hall and for parlor one with an area of 75 square inches.

Hot-Air Pipes. Run a tin pipe for hot air from indirect stacks to connect with registers with an area of 180 square inches for main hall and one with an area of 100 square inches for parlor.

Dampers for Cold-Air Pipe. Place in each cold-air pipe for direct-indirect radiators a damper beneath radiator, with handle extending outside radiator at end; also dampers in each cold-air pipe for indirect stacks same size as pipe.

Registers. Place in floor above indirect stacks where marked on plans one register for main hall 14 x 22 inches, and for parlor 12 x 14 inches, to be electro-plated and with borders for same; also place in ceiling of piazza for parlor chamber a 7-inch round register face, and for library chamber a 6-inch round register face, and for dining-room chamber 7½-inch round register face, and for smoking-room 5½-inch round register face ; these to be finished in white enamel. Where cold-air pipes for indirect stacks enter building place a register face 14 x 16 inches for the main-hall pipe and 10 x 12 inches for the parlor; these to be finished in plain black.

Mains. Mains to be of number and size marked on plans to supply all radiators; all pipe used to be of best quality and standard make. All over 1¼ inches to be lap-welded. All radiators on first floor up to 40 square feet to have 1-inch flow and return connections, and all larger than 40 square feet to have 1¼-inch flow and return connections. On upper floors up to 50 square feet to have 1-inch flow and return; than larger 50 square feet to have 1¼-inch flow and return connections Horizontal flow and return mains in cellar to be suspended from joists in ceiling, to rise 1 inch in 10 feet from heater to radiators. All rising pipes to be plumb and straight.

Supply to Indirects. Main feed-pipes for indirect radiators to be carried up to second-floor joists and turn down to feed radiators, and the return to run under cellar floor in a pine-board box 1¼ inch thick; cover of same to be left loose and to be level with cellar floor. From top of main feed-pipe carry a ¾-inch air-pipe to connect with overflow of expansion above water-line.

Circulation. All connections from mains to the branches and from the branches to the radiators are to be so made that a free and continuous circulation of water in the whole apparatus will be obtained at all times and under all circumstances when apparatus is in use.

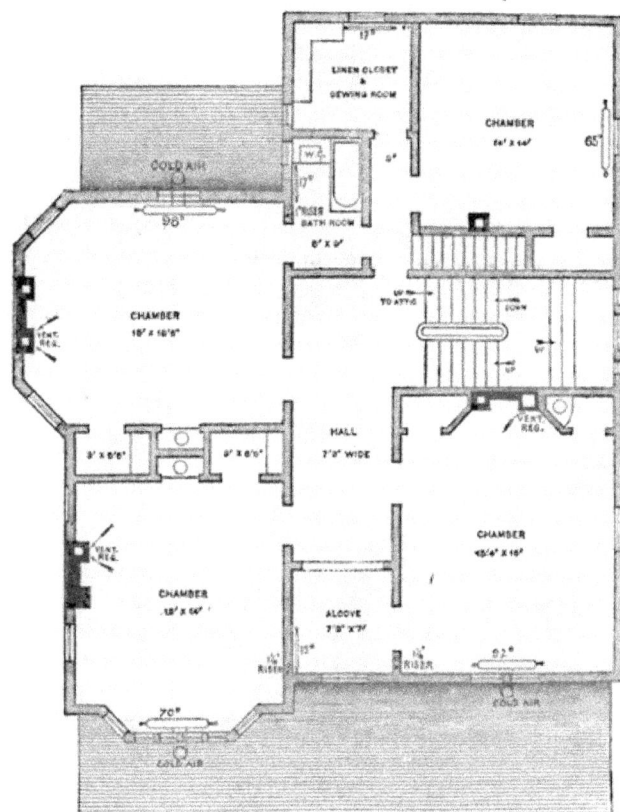

Fig. 3.—Second-Floor Plan.—Scale, 1-12 Inch to the Foot.

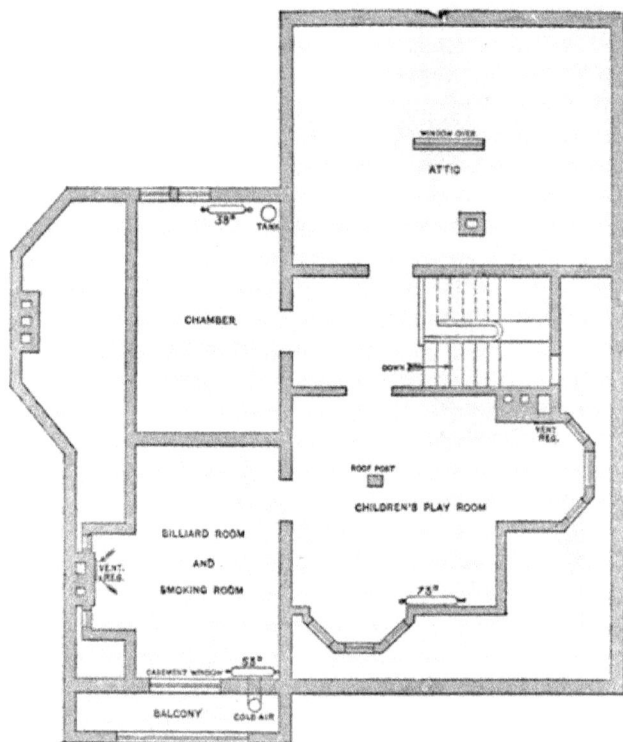

Fig. 4.—Attic Plan.—Scale, 1-12 Inch to the Foot.

All fittings to be heavy-pattern gray cast-iron; no malle- Fittings. able iron or union couplings to be used in the work. The elbows on radiators on the opposite end to the valves to be nickel-plated, and floor-plates around pipes coming through floors to radiators to be nickel-plated.

Where pipes pass through walls, floors or ceilings the Floor and Ceiling Plates. opening to be covered with cast-iron floor and ceiling plates properly secured in place; plates to have sleeves to extend through floors and ceilings.

Where pipes pass through brick walls in cellar to have cast- Sleeves. iron sleeves built in wall, to be larger than outside diameter of pipes, for running pipe through.

Horizontal mains in cellar to be supported by cast-iron Pipe Hangers. ring hangers 8 feet apart properly secured to joists in ceiling.

The ends of all pipes used in the work to be reamed out Reaming. and all the burrs and obstructions removed before being placed in position.

Place a 1-inch brass stop-cock at lowest point in heater ; Draw-off Cock & Pipe. connect same with 1-inch iron pipe to drain.

Place in cupboard at end of pantry sink a coil of 1-inch Plate-Warmer. pipe, containing not less than 60 feet of 1-inch iron pipe, to be arranged in shelves for use as a plate-warmer, the doors in the front of sink to have registers in same; so that coil will heat pantry.

Each radiator to have an angle-valve, full opening, of Radiator-Valves. heavy pattern on flow main, full size of same, valves to have wood wheels and polished bodies, nickel-plated all over.

Each radiator to have a nickle-plated air-valve at highest Air-Valves. point, air-valves to be fitted for keys; furnish two keys for same.

Place gate-valves of heavy pattern on each flow and re- Valves for Mains. turn main of full size of same at boiler, with ½-inch draw-off cocks on side away from boiler for the purpose of drawing off water from same, should it become necessary.

Place in attic chamber, on wood shelf supported by orna- Expansion-Tank. mental iron brackets, one closed-top galvanized-iron expansion tank, constructed of No. 10 gauge iron, 2 feet long, to hold 18 gallons, to be fitted with gauge glass and fixtures; also self-closing cock and funnel. Connect tank to water-supply of house with ½-inch iron-pipe and ½-inch brass stop-cock, a check-valve to be placed between cock and tank. Run 1-inch iron pipe for overflow from top of tank to laundry sink. Where overflow-pipe turns down place 1-inch vacuum-valve. Run 1-inch supply-pipe from bottom of tank to connect on return-pipe inside main valves at boiler.

Marbles for Radiators. Place under each radiator on first floor 1¼-inch marble slabs counter-sunk in center ¼ inch and to extend outside pipes coming through at least 1½ inches. Place on second and third floors where direct-indirect are not used 1¼-inch walnut boards to take place of marble slabs; also furnish marble slabs for tops of radiators. All marbles to be of desired colors.

Indicator. Place a hot-water indicator on heater where it can easily be seen, for the purpose of ascertaining the temperature of the water.

Bronzing. All radiators and exposed pipes to be neatly painted and bronzed in desired colors; also mains and heater in cellar to have two coats of black japan.

Ventilating-Registers. Furnish registers for ventilation of desired finish and sizes as follows: Parlor, two 11 x 14 inches. Dining-room, two 10 x 14 inches. Library, two 10 x 14 inches. Parlor chamber, two 10 x 12 inches. Dining-room chamber, two 10 x 12 inches. Library-chamber, two 10 x 10 inches. Smoking-room, two 8 x 10 inches. Children's play-room, two 10 x 10 inches. Bath-room, two 6 x 8 inches.

Toilet-room, first floor, two 6 x 6 inches. Registers to be set in flues which are to be provided by owner and of size named as follows:

Ventilating-Flues. Parlor flue to have an internal area of 90 square inches. Library, 84 square inches. Parlor chamber, 70 square inches. Dining-room, 84 square inches. Dining-room chamber, 78 square inches. Library-chamber, 60 square inches. Smoking-room, 48 square inches. Children's play-room, 63 square inches. Flues to be independent of each other and to extend to floors of each room, so that bottom register can be placed as low as possible. Run tin pipe for bath-room and toilet from floor of same up through roof; for bath-room of an internal area of 25 square inches; for toilet-room with an internal area of 20 square inches. All flues with the exception of bath-room and toilet to be provided by owner.

Boxing for Indirects. Indirect radiators in cellar to be boxed in pine boards 1 inch thick, tongued and grooved, neatly made and lined with tin.

Proposal and Estimate. I propose to erect an apparatus of this kind in accordance with the foregoing plans and specifications, temperature to be in all rooms of first floor 70° in zero weather, halls and chambers 65°, and bath-room 75°, smoking-room and children's play-room, 70°; all materials to be as specified; the work to be completed in a good and workmanlike manner and to the entire satisfaction and acceptance to owner, for the sum of $1725.

ESTIMATE.

592 square feet radiators, 24 inches high, @ 48 .... $284.16

301 square feet radiators, 36 inches high, @ 40 .... 120.40

280 square feet radiators, indirect, @ 30 .... 84.00

Boiler and fire-irons 250.00

Smoke-pipe 5.00

Bronzing and varnishing 25.00

Valves and air-valves 154.00

Mains and fittings 150.00

Boxing indirect stacks........ ... $30.00

Hot and cold air-pipes......... 25.00

Registers.................... 52.50

Labor, eight weeks for man and help.... .... 280.00

Marble slabs, &c.............. 200.00

Cartage and freight and fares... 50.00

Sundries..................... 15.00

Total..................... $1725.06

THE STEAM AND HOT WATER HEATING COMPETITIONS.*

One of the experts who assisted the judges in *The Metal Worker* Steam and Hot Water House Heating Competitions has made the following summary of the several essays, and compared in tabular form the methods of heating, the amounts of heating surface, its distribution, &c. It is not to be supposed that the averages and

Floors.	Number of apartments, halls, &c., on each floor.	Space on each floor in cubic feet.	Aspect.	Square feet of glass.	Total square feet of glass on each floor.	Square feet of exposed wall.	Total square feet of exposed wall on each floor.	Total equivalent of wall to glass at 10 to 1 inch square feet.	Total square feet equivalent to glass.	Exposed wall in lineal feet.	Total lineal feet of exposed wall on each floor.	Percentage of cubic space on each floor.	Percentage of equivalent to glass on each floor.	Percentage of lineal feet of exposed wall on each floor.
Third...	4	6189	North.	10.5	.....	91.5	.....	.....	.....	15.5	.....	....	....	...
			South.	5.0	.....	116.0	.....	.....	.....	25.0	.....	....	....	...
			East.	14.5	....	89.0	.....	.....	.....	22.0	.....	....	....	...
			West.	8.0	38.0	80.5	377.0	77.7	75.7	10.0	72.5	17	9	20
Second.	8	14082	North.	42.5	.....	391.5	.....	.....	.....	41.5	.....	....	....	...
			South.	45.5	.....	296.5	.....	.....	.....	31.0	.....	....	....	...
			East.	65.0	.....	361.0	.....	.....	.....	38.0	.....	....	....	...
			West.	25.5	178.5	322.0	1371.0	137.1	315.6	34.0	144.5	38	35	38
First...	10	15909	North.	66.5	.....	451.5	.....	.....	.....	43.9	.....	....	....	.
			South.	62.0	.....	388.5	.....	.....	.....	37.0	.....	....	....	.
			East.	147.0	.....	433.0	.....	.....	.....	41.0	.....	....	....	...
			West.	48.0	323.5	370.0	1643.0	164.3	487.8	35.0	156.0	45	56	42
Totals..	22	36180	...	.....	540.0	.....	3391.0	339.1	879.1	....	373.0	.....	....	....

Table I.—Summary of Space, Surfaces, &c.

deductions given necessarily represent the best practice, for in such work as steam and hot water heating there are no absolute standards formulated as yet. The tables and text are, however, both interesting and valuable, for they present in a concise form the practice of many different experts engaged in the practical work of steam and hot water heating in various parts of this country and Canada.

* Reprinted from *The Metal Worker*, March 8, 1890.

Table I gives the general information necessary in order to draw up a specification and estimate for heating, by steam or water, the house described in the prospectus of *The Metal Worker* competition. Tables II and IV give various methods of proportioning and ascertaining the quantity of radiating surface, and its distribution on the different floors ; while Table III presents a summary of the estimates. For convenience in comparison we have classified the com-

Methods of Heating.	Number of square feet of surface in radiators to			Number of cubic feet of space to 1 square foot of surface in radiators.	Distribution of surface in radiators on floors.		
	1 square foot of glass, or its equivalent.	1 lineal foot of exposed wall.	100 cubic feet of space.		Floors.	Percentage of direct radiating surface.	Percentage of indirect radiating surface.
Direct radiation only............	.469	1.10	1.136	88.	1st	65	...
					2d	24	....
					3d	11	....
Direct-indirect radiation only	.600	1.41	1.470	68.	1st	59½	....
					2d	81½	....
					3d	9	....
Direct-indirect and direct radiation, 65 per cent. of the surface being direct-indirect	.627	1.48	1.538	65.	1st	52	...
					2d	35½	....
					3d	12½	....
Indirect and direct radiation, 87½ per cent. being direct radiation.	.670	1.59	1.666	60.	1st	35.6	13.5
					2d	31.8	....
					3d	19.7	....
Indirect and direct radiation, 45½ per cent. being direct radiation....	.855	2.02	2.083	48.	1st	3.75	56.5
					2d	28.25	....
					3d	13.5	....
Indirect and direct radiation, 55 per cent. being direct radiation..	.921	2 17	2.222	45.	1st	14.37	31 45
					2d	29.87	9.27
					3d	10.76	4.28
Indirect radiation only..	.771	1.85	1.923	52.	Not capable of averaging.		

Table II.—Steam Radiating Surfaces.

petitors who agree on the several points according to percentages of the whole number.

STEAM HEATING COMPETITION.

The Method of Heating Used :

Direct radiation only, 11 per cent.
Direct-indirect radiation, 22 per cent.
Direct and indirect radiation, 55 per cent.
Indirect radiation only, 11 per cent.

Number of Apartments and Halls Heated :
   14 apartments were heated by 22 per cent.
   15 apartments were heated by 33 per cent.
   16 apartments were heated by 22 per cent.
   18 apartments were heated by 22 per cent.
   Hall on third floor was heated by 11 per cent., while the hall on second floor was heated by 44 per cent. and the vestibule on first floor was heated by 22 per cent.
   Hall on first floor was heated by direct radiation only by 33 per cent., by indirect radiation only by 55 per cent. and by direct-indirect by 11 per cent.; 22 per cent. heated the butler's pantry, half of the number using indirect and the others direct radiation.
   The kitchen was heated by direct radiation by 44 per cent.

From Table II the average distribution of radiating surface to each floor is 53.7 per cent. to the first floor, 32.59 per cent. to the

Cost of pipes and fittings.	$180 to $185	$120 to $125	$60 to $80	$50 to $55	.	
Percentage...	11	22	55	11		...
Cost of bronzing and decorating radiators and pipes...	$18 to $20	$12 to $13	$5 to $7	....	....	
Percentage	22	11	22	....	....	..
Cost of covering pipes in cellar	$110 to $115	$50 to $55	$20 to $30	$10 to $15	....	
Percentage...	11	11	44	11	......	..
Cost of labor, pipe fitting.	$310 to $320	$130 to $140	$100 to $110	$90 to $99	$80 to $89	$50 to $55
Percentage...	11	11	22	33	11	11
Total estimates.	$1600 to $1800	$1300 to $1400	$900 to $1000	$800 to $899	$700 to $789	$600 to $699
Percentage.	22	11	33	11	11	11

Table III.—Comparison of Estimates.

second floor and 13.71 per cent. to the third floor, the largest proportion of surface on the first floor being 65 per cent. and the least or lowest proportion being 43.30 per cent.; to the second floor 41.15 per cent. being the highest and 24 per cent. the lowest, and to the third floor the greatest proportion given is 19.70 per cent. and the least 9 per cent.

In all the specifications fresh air inlets were provided. While 78 per cent. provided exits for foul air, 22 per cent. omitted any mention

of these exits.   The areas in square inches of the fresh air inlets may
be thus given:

11 per cent. provided an average area of from 1900 to 2100 square inches.
11 per cent. provided an average area of from 1200 to 1400 square inches.
44 per cent. provided an average area of from 600 to 800 square inches.
11 per cent. provided an average area of from 400 to 500 square inches.
22 per cent. provided an average area of from 30 to 200 square inches.

The areas of the foul-air exits may also be summarized:

22 per cent. provided an average area of from 800 to 100 square inches.
33 per cent. provided an average area of from 400 to 600 square inches.
22 per cent. provided an average area of from 150 to 300 square inches.

In the fresh air inlets 44 per cent. averaged about 50 cubic feet
of space to 1 square inch in the foul air exits to about 70 cubic feet of
space in the average of 33 per cent. of all the proposals, or about 43
per cent. of those who gave sizes of foul air exits.

Cast iron radiators are apparently more generally used than
wrought iron radiators, as 66 per cent. used the former, while only 22
per cent. adopted the latter and 11 per cent. offered pipe coils.   Forty-
four per cent. used 1¼-inch steam and 1-inch return valves on all
radiators, direct or indirect, while 22 per cent. used 1-inch valves on
steam and ¾-inch valves on return on direct radiators having less
than 50 square feet of surface.   Forty-four per cent. gave no sign of
steam or return valves on radiators.

The system of piping most generally adopted was that known as
the double pipe—that is, steam and return pipes from all radiators ;
55 per cent. used this method, 33 per cent. used the same method to
the first and the single pipe system to the second and third floors—
that is, one pipe instead of two to each radiator—and 11 per cent.
used the single pipe system throughout to all radiators, with main
steam and return pipes in the cellar ; 11 per cent. used 1¼-inch pipes
to the radiators in the single pipe system, and the same percentage
offered 2-inch pipes in the same system.

The main steam pipes were of various sizes ; 22 per cent. used
one 3-inch main, the same percentage had two mains each of 2½-inch
pipe ; 11 per cent. used one 3½-inch pipe and the same proportions
had respectively two mains, one 3-inch pipe and one 2½, and the
other one 2¼-inch pipe and 2-inch pipe.   The remaining 33 gave no
definite information relative to the main steam pipes.

A 2-inch main return was used by 55 per cent. of the competitors,
one or two others used 2½-inch return main, and the others gave no
definite information about the sizes of return mains ; 88 per cent.

placed the main return below water line in boiler. Check valves on return pipes were used by 22 per cent., while only one provided any means to prevent water backing up in radiators.

The position and style of finishing connections were only referred to in a casual way; 22 per cent. mentioned the use of flange

Methods of heating.	Number of methods compared.	Number of square feet of surface in radiators to			Number of cubic feet of space to 1 square foot of surface in radiators.	Distribution of surface in radiators on floors.				
		1 square foot of glass, or its equivalent.	1 lineal foot of exposed wall.	100 cubic feet of space.		Floors.	Percentage of direct radiating surface.	Percentage of indirect radiating surface.	Percentage of direct-indirect radiating surface.	Total percentage.
1.—Direct radiation...	20%	0.986	2.88	2 455	4.	1 / 2 / 3	45 / 35 / 20	....	....	
2.—Direct, direct-indirect and indirect radiation......	10%	1.651	3.190	3.256	30	1 / 2 / 3	31.5 / 9.5 / 7.5	23.5 / .... /	.... / 29 / 6	55 / 31.5 / 13.5
Totals........... .							48.5	23.5	28	
3.—Direct and indirect radiation	70%	1.243	2.938	3.030	33	1 / 2 / 3	23.3 / 26.8 / 13.7	29.7 / 4.2 / 2.3	....	53 / 31 / 16
Totals.............							63.8	36.2		
4.—Sub-division of No. 3 comparison of direct and indirect radiation	49%	1.121	2.650	2.732	36	1 / 2 / 3	39 / 35.5 / 16.5	9 / .... /	....	48 / 35.5 / 16.5
							91	9		.... ..
	17%	1.113	2.631	2.792	37	1 / 2 / 3	.... / 24 / 15.2	60.8 / .... /	....	60.8 / 24 / 15.2
							39.2	60.8		
	34%	1.483	3.005	3.570	28	1 / 2 / 3	28.5 / 22 / 10.7	22.5 / 10.5 / 5.8	....	51 / 32.5 / 16.5
							61.2	38.8		....

Table IV.—Hot Water Radiating Surfaces.

or other unions on main pipes, and only 11 per cent. specified ground unions or radiator valves.

Boilers may be enumerated as follows: Wrought iron boilers were used by 55 per cent., cast iron boilers were used by 33 per cent. and 11 per cent. did not state of what material the boilers were to be

made; 44 per cent. had brick set boilers, while 55 per cent. used those
of the portable type.   Magazine coal feeders were provided with 44
per cent. of the boilers.   The following proportions are of interest.

Grate surface provided ·
6 and above 5 square feet, by...	22 per cent.
5 " " 4 "	44 "
4 " " 3 "	33 ·'

Boiler heating surface :
200 to 210 square feet, by.....	11 "
160 to 170 "	11 "
140 to 150 ·'	33 "
110 to 120 "	22 "
90 to 100 "	22 "

Square feet of boiler surface to 1 of grate surface :
40 to 50...	11
30 to 39...	22 "
20 to 29...	66 "

Square feet of radiating surface to 1 of boiler surface :
6 and above 5...	44 '·
5 " " 4..................................	44 "
4 " " 3...	11 "

The use of automatic water feeders was disapproved of by 22 per
cent., and the same proportion of competitors did not mention them,
while 55 per cent. specified and used them ; 66 per cent. used automatic damper regulators and the others did not refer to them.   Electrical thermostatic regulators were specified and used in addition to
the common regulator by 55 per cent.

Hardwood boards or marble slabs under radiators were only specified by 11 per cent. of the competitors, and automatic air valves
with drip pipes were specified by 55 per cent., the others using hand
air valves or not definitely explaining what they did propose to use,
or making no reference whatever to them.

The remaining items in these specifications which attract attention are the varieties of the estimated costs and the differences in the
total estimates.   The value of the pipes and fittings is variously estimated, the lowest being $50 and the highest about $185.   The cost of
decorating and bronzing radiators and pipes is mentioned as from $5
to $20, a proportion of 44 per cent. giving no price.   The lowest sum
put down for pipe covering is $10 and the highest is $115, and 22
per cent. give no estimate on this item.   The cost of the labor connected with the pipe fitting varies from about $50 to $320, and the
total estimated value of the job goes from about $600 to $1800.   In
Table III the estimates are compared in more detail.

In comparing the estimated costs, it will not be possible to make detailed comparisons in all or the principal items, several having omitted detail prices and included different articles. From the separate prices given for the expansion tank, it is estimated as worth from $5 to $15.

The smoke pipe cost is placed as low as $2 by one and as high as $10 by others. The estimates of the cost of radiation vary by those who give details from $227 to $570, while an average of all the prices given makes its value about $380. Pipes and fittings range from $43 to $415, but it is possible something else is included in the highest one, because without it the average cost is $121. The estimated time required to do the pipe fitting is placed at from 5 to 48 days ; an average of all the times proposed gives 21 days, while the cost is stated to be from $21 to $280, and when all the costs are taken an average of $121 is the result. An item in estimating which was very generally omitted was the cost of freight ; only a few mentioned it, and it may be averaged at about $35. The total estimates varied from $775 to $1725, and the average was $995, which is probably a fair sum at which to place the cost of heating such a residence by hot water.

HOT-WATER HEATING COMPETITION.

Methods of Heating Used :

    Direct radiation only, 20 per cent.
    Direct and direct indirect and indirect radiation, 10 per cent.
    Direct and indirect radiation, 70 per cent.

Number of Apartments and Halls Heated :

    11 apartments were heated by 20 per cent.
    14 apartments were heated by 10 per cent.
    16 apartments were heated by 20 per cent.
    17 apartments were heated by 10 per cent.
    18 apartments were heated by 10 per cent.
    19 apartments were heated by 10 per cent.
    21 apartments were heated by 20 per cent.
    Hall on third floor was heated by direct radiation by 40 per cent.
    Hall on second floor was heated by direct radiation by 70 per cent.
    Vestibule was heated by indirect radiation by 20 per cent.
    Main hall, first floor, by direct radiation only, 50 per cent.
    Main hall, first floor, by indirect radiation only, 20 per cent.
    Main hall, first floor, by direct and by indirect radiation, 20 per cent.
    The back hall, first floor, was heated by direct radiation by 30 per cent. and by indirect radiation by 10 per cent. ; 40 per cent. heated butler's pantry by direct radiation and 10 per cent. by indirect radiation. The kitchen is heated by direct radiation by 30 per cent. and by indirect by 10 per cent.

From Table IV it will be ascertained that when direct radiation is used 45 per cent. of the surface is placed on the first floor, 35 per cent. on the second and 20 per cent. on the third floor, whereas, where indirect and direct radiation is used the surface is proportioned to the floors respectively 53, 31 and 16 per cent.

Fresh air inlets were provided by 80 per cent. of the competitors, and they gave the number of square inches in these openings: 50 per cent. had a total area of 520 square inches and 30 per cent. for the same had 163 square inches, making an average of 341 square inches. Only 30 per cent. gave the areas of the fresh air openings to each floor, and these average 192 square inches to the first floor, 97 to the second floor and 44 to the third floor.

Foul air exits were specified by 50 per cent., some of the others referred to chimney flues as sufficient, while others made no mention of them. The average of the total areas of these exits is 500 square inches.

Cast iron radiators were specified by 90 per cent. of the contestants, 10 per cent. using 1-inch pipe coils; 20 per cent. used indirect pin radiators; 60 per cent. provided coils of 1-inch pipe for the indirect radiators; 20 per cent. gave no sizes of the flow and return connections to radiators; 30 per cent. used 1½ and 1¼ to indirect and first floor radiators and 1 inch to second and third floor radiators; 20 per cent. had 1¼, 1 and ¾-inch flow and return on first, second and third floors; 10 per cent. had 1-inch connections on all floors and about the same number had 1-inch connections to first floor and ¾-inch to third floor.

The materials used in the construction of the boilers or heaters were not in all cases described, and some omitted the size of grate. Forty per cent. specified cast iron and 20 per cent. wrought iron tubes in the boilers.

Grate surface provided :
    6 square feet and above 5 square feet, by.................. 10 per cent.
    4 square feet.. 40   "
    3 square feet and under 4 square feet.................... 30   "
Boiler heating surface :
    110 square feet to 120 square feet....................... 20   "
    80 square feet... 40   "
    60 square feet and under 80 square feet.................. 10   "
Boiler heating surface to 1 square foot of grate :
    20 to 1... 40   "
    15 and under 20 to 1..................................... 10   "
    30   "   40 to 1····.......... 30   "

Square feet of surface in radiators to 1 square foot of boiler
    heating surface :
    7 to 1 and under 10 to 1............ 10   "
    10 to 1   "   12 to 1.......................... 10   "
    13 to 1   "   15 to 1............................ .. 20   "
    15 to 1   "   17 to 1............................ 10   "
    17 to 1.. 10   "

The main flow pipes can only be compared by the number and a general reference to the sizes of pipes. Seven flow pipes were specified in 10 per cent. of the specifications, the largest and smallest of these pipes being respectively 2½ and 2 inches ; 40 per cent. recommended six flow pipes, the largest being of 2½-inch pipe and the smallest 1¼-inch pipe ; 20 per cent. proposed five flow pipes, from 2½ to 1½ inches being the sizes, and 20 per cent. had four flow pipes with 3½ to 2 inch pipes as the largest and smallest pipes used. Some used single mains, but the information is not sufficient to make a comparison of any value.

Marble slabs under radiators were specified by some 20 per cent. of the competitors, while thermostatic and automatic regulators were used by about the same number. The covering of the main pipes was specified by 90 per cent. While 60 per cent. connected the expansion tank to bottom of return pipe near boiler, 40 per cent. connected to the return pipe of a radiator.

IV. HOT-AIR SYSTEMS.

HOT-AIR SYSTEM.*

BY ANSON W. BURCHARD.

SPECIFICATION.

The importance of ventilation is universally acknowl- Ventilation. edged, and the connection of the heating of a house with its ventilation is so inseparable that no heating apparatus which does not combine with it as thorough a system of ventilation as practicable can be considered complete.

Every one who has had occasion to examine the subject knows that very few buildings are provided with efficient means of ventilation, and that however well the heating apparatus may be calculated to maintain the temperature at the desired degree in the coldest weather in very few cases does it insure in connection with this an abundant supply of fresh air. Where the question of expense and attendance does not enter the problem, to secure this supply of fresh air is not such a difficult matter, and many large buildings are fitted with appliances for this purpose of a very complete description. Such apparatus require the services of an attendant almost constantly, and are not, therefore, practicable for use in a private residence such as the one under consideration.

Hence, in designing an apparatus for such a house, any ventilating appliances that are adopted should be automatic and the movement of the air induced by the natural drafts of chimneys and hot flues, fans and other mechanical devices being impracticable because of the attention required to keep them in operation. In arranging appliances to afford this ventilation, one of the first points to be considered is how much fresh air will be furnished.

Perfect ventilation may be said to have been secured in an inhabited room only when any and every person in that room takes into his lungs at each respiration air of the same composition as that surrounding the building and no part of which has recently been in his own lungs or those of his neighbors, or consists of products of combustion generated in the building, while at the same time he feels no currents nor

* From *The Metal Worker*, September 7, 1889. Copyrighted, 1889, by David Williams.

drafts of air and is perfectly comfortable as regards the temperature, being neither too hot nor too cold.

Very rarely can such ventilation be secured if the number of occupants of a room exceeds two or three.

Air Impurities. Without entering into a discussion of the methods and expense of securing perfect ventilation, good ordinary ventilation is to be secured by keeping the vitiated air diluted to a certain standard. All air with which ventilating appliances have to deal contains more or less impurities, some of which are more dangerous than others and are less affected by this process of dilution. Of these impurities carbonic-acid gas is popularly supposed to be the most harmful, but, as a matter of fact, it is not poisonous, and produces no harmful effect even when present in 30 to 50 times the normal quantity. But this carbonic acid is generally found accompanied by other gases which are harmful, particularly carbonic oxide and sulphureted hydrogen. Hence, as there is no convenient method of determining the percentage in which the two latter gases are present, it is usual to determine the percentage of carbonic acid, for which there is a simple method, and assume that the amounts of the other gases present are proportionate to this.

The Odor Test. As a rule an apartment may be considered well ventilated when a person entering it from the fresh outer air does not perceive any special odor ; and experience has shown that a faint, musty, unpleasant odor is perceptible under such circumstances if the amount of carbonic acid present, of which the normal is about 4 parts in 10,000, be increased to above 7 parts in 10,000. If the air which has been used and contaminated did not mix with the air in the room a comparatively small amount of fresh air would be required to keep this up to the standard of purity. Basing their estimates on this **Fresh-Air Supply.** erroneous assumption, some authorities have concluded that 250 cubic feet of fresh air per hour for each occupant is all that would be required ; but as the contaminated air does mix with the fresh air it is found that in order to keep the carbonic acid diluted to 7 parts in 10,000 of air a supply of 3000 cubic feet of fresh air per hour is needed for each occupant where rooms are continuously occupied.

Capacity of Heating Apparatus. The rooms of dwelling-houses are rarely occupied for a long time continuously, and there is a considerable amount of air admitted through the accidental openings ; besides, the amount of cubic space (breathing space) *per capita* is very large, so that a heating apparatus supplying 2000 cubic feet of fresh air per hour for each occupant answers every requirement for a dwelling-house.

Room.	Area of air-pipe, square inches.	Supply of air, cubic feet per hour.
Parlor.........................	154	10,500
Hall...........	113	8,400
Library.........................	113	8,400
Dining-room.	113	8,400
Lavatory	31¼	2,400

Thus the parlor would have a supply adequate for five persons, the hall, library and dining-room each for four, assuming the registers to be wide open and enough fire in the furnace to create the assumed velocity of air movement in the flues. It is proposed to provide in the parlor, library and dining-room ceiling ventilators with separate flues for each for the purpose of carrying off the products of combustion of the gas without allowing them to mix with the air of the room to any great extent. Ceiling Ventilators.

The movement of the air in the flues to the second and third floors would probably not be less than 5½ feet per second. This would give the chamber on the north side a supply of 7000 cubic feet per hour and the two chambers on the south side a supply of 7000 cubic feet together—that is, 3500 feet each.

As these rooms would not be occupied by a large number of people for any length of time continuously, this supply of air would be abundant. The air which would have to be introduced into the other rooms to keep them warm enough would be enough to secure good ventilation for any purposes for which they would be likely to be used. But the air is not delivered to the rooms at the same temperature at all times. In cold weather the air must be hotter to keep the rooms at the proper temperature, and in warm weather cooler.

As the air-movement is due to the expansion of the air by the heat, it follows that the higher the degree to which the air is warmed the greater the quantity which will flow through the pipes and enter the rooms in a given length of time.

The velocities assumed in the above estimates would not be maintained unless the air were heated to a degree above that required to keep the room warm in mild weather—that is to say, in mild weather the supply of heat and consequently air must be reduced or the rooms will be overheated. In some forms of apparatus this is in a measure overcome by mixing cold air with the warm air in mild weather, and thus reducing its temperature before it enters the room. But with these appliances, called mixing-dampers, the velocity and consequently the volume of the air is lessened because Variable Supply.

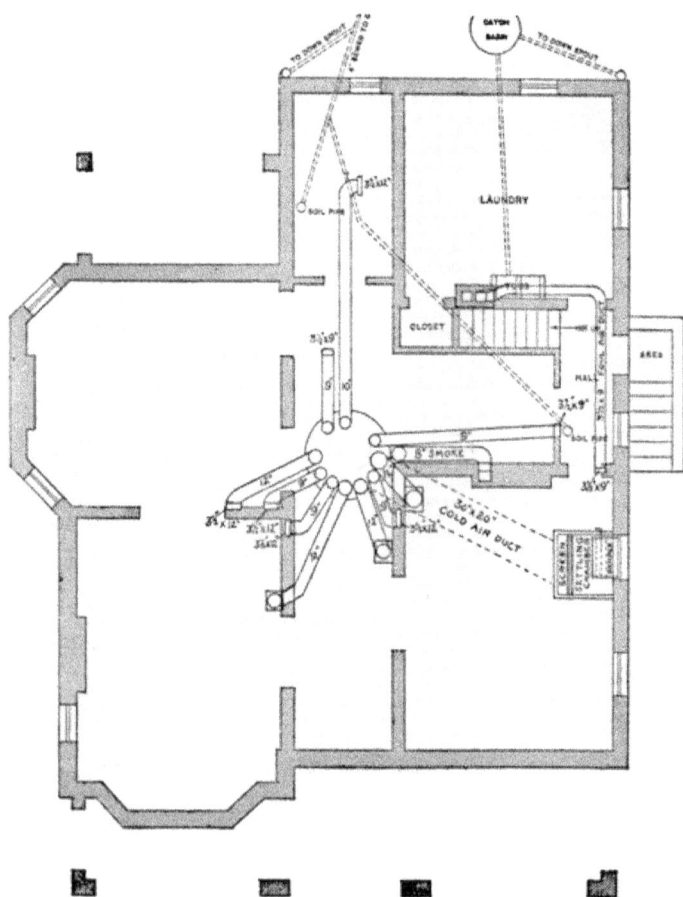

Fig. 1.—Cellar Plan.—Scale, 1-12 Inch to the Foot.

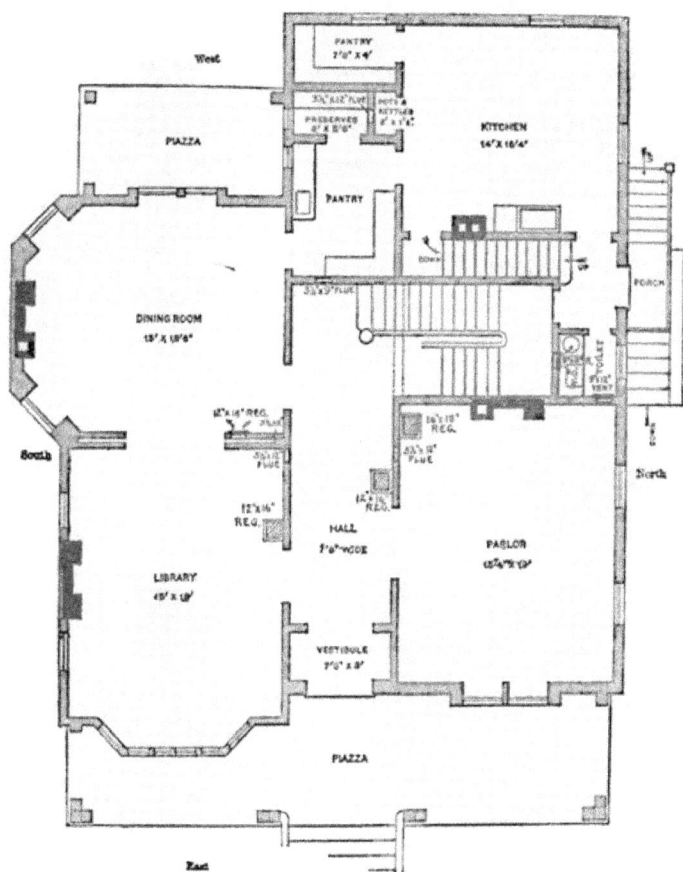

Fig. 2.—First-Floor Plan.—Scale, 1-12 Inch to the Foot.

of its reduced temperature, and they do not operate successfully in connection with a system of heating which has but
one out-of-door air-supply for all the rooms, as is a necessity
with a hot-air furnace.

Also, when the gas is lighted in a room a considerable
amount of heat is given off, so that the supply of heat, and
consequently air, from the furnace must be reduced to compensate for this, otherwise the room becomes overheated.
This is why the want of proper ventilation is more frequently
noticed in the evening.

Large Flues. These difficulties can be largely overcome in heating a
house with a hot-air-furnace by providing flues of large capacity and a furnace of large radiating-surface, which does
not require to be raised to a high temperature to give off the
required amount of heat, and by these means insuring the
introduction into the rooms of a large volume of moderately-
heated air.

Large Furnace. For this reason the furnace which is proposed in this design is very much larger than in the ordinary practice of
furnace-setters would be used for a house of this size.

Description of Furnace. The furnace which is proposed is made almost entirely of
cast-iron. It is assumed that the castings would be made
of a good quality of iron by foundry men of intelligence and
experience, who would insure castings sound and free from
sand-holes.

Dimensions. If a proper amount of care be taken, there is no difficulty
in producing such castings. The fire-pot would be 35 inches,
inside diameter, 15 inches deep, and 1½ inches thick. Sur-
mounting this fire-pot is a dome 32 inches high, with a round
top. This dome has four hollow arms extending out from a
line a little above the line of connection between it and the
fire-pot, and equidistant from one another. Surrounding
this dome and connected with its arms by corresponding collars is an annular radiator, 55 inches in diameter.

The fire-pot sets on a cast-iron base, forming the ash-pit,
which is 18 inches deep.

Grate. The grate is composed of four heavy cast-iron pieces, arranged parallel to one another and running from front to
rear. The cross-section of these pieces is approximately an
equilateral triangle with slightly-concaved sides. These
pieces have at each end a journal which rests in suitable
bearings at the front and rear of the ash-pit, and which are
so located that when these pieces are in position they will revolve freely, with a clear space of about 1½ inches between
one another. Keyed on to the front end of each of these
pieces is a gear-wheel. These wheels attached to the two

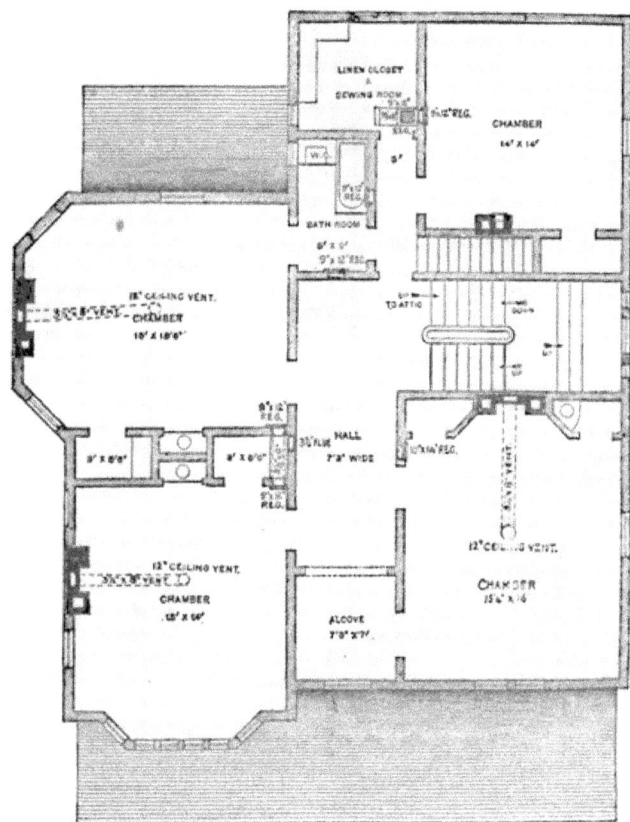

Fig. 3.—Second-Floor Plan.—Scale, 1-12 Inch to the Foot.

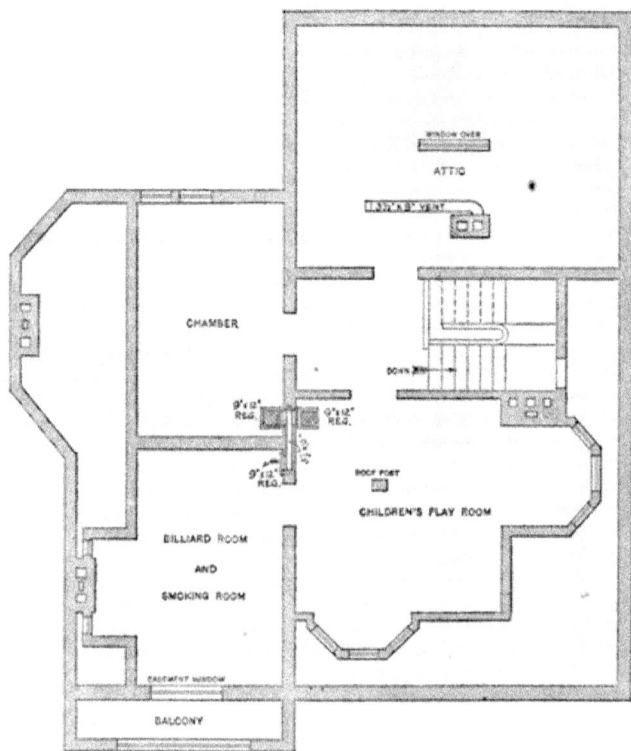

Fig. 4.—Attic Plan.—Scale, 1-12 Inch to the Foot.

middle pieces each engage the wheel on the outside piece
next to themselves, so that if a wrench or "shaker" be attached and one of the middle pieces revolved the outside
piece will revolve correspondingly in the opposite direction.
Thus the two halves of the grate can be shaken separately,
and very little effort is required to do this, which is of
special advantage where the grate is of large area.

With this form of grate a slight oscillation of the shaker
is all that is required to free the grate from an accumulation
of ashes, and a part of a revolution will remove all the clinkers. This grate also has the advantage that it cleans the fire
equally all over, which grates with a movement in a horizontal plane do not do, and this keeps an even fire all over, preventing the formation of a troublesome mass of clinker in
the center.

The base of the furnace is made in two pieces, which are
well fastened together, forming an ash-pit of good depth with
plenty of air-space underneath the grates as a precaution
against their burning out. The door to this ash-pit opens
the full depth. The usual dust-flues, dampers, sifting-grate,
&c., if desired are provided.

The base has at the top on each side near the back a
strong lug, to which are bolted strong legs, which in turn
rest on the brick-work forming the sides of the furnace-pit,
and support the weight of the furnace. This method of support is preferable to the common method of resting the base
on a brick pier, as it allows the base to expand and contract
freely, permits the air to have access to all parts of it, and
does not obstruct the passage of the air through the furnace
by a pier; all of which tend to prevent warping and cracking of the castings.

The lower edge of the fire-pot sets into a deep cup on the
top of the base. This joint is made tight by spreading in
the bottom of the cup a thin layer of furnace cement before
the fire-pot is set on, and after this is in place filling the cup
up with fine sand. The dome sets into a similar cup on top
of the fire-pot, and the radiator into similar cups on arms of
the dome, all of which joints are made tight in the same
manner.

The dome is not less than ½ inch thick in any part, and
the radiator not less than ⅜ inch. The radiator is divided
by a horizontal diaphragm into an upper and lower chamber,
which are connected together on the side opposite where
the smoke-pipe is taken off.

The furnace has what is called a "panel" front, which
is a large casting about 36 inches wide and the full hight

of the furnace. Through this the openings are made, and to it the doors are hinged and the casings bolted. There are four of these doors, a feed-door with a perforated cast-iron lining, two cleaning-doors, with tin linings, and an ash-pit-door. The feed-door has the usual damper, and the ash-pit-door a sliding damper, and a balanced draft-door for attaching the automatic regulator.

Jacket.   The furnace is cased with a jacket of No. 22 galvanized iron, which is lined with corrugated tin, fastened to the iron with rivets. Inside the jacket, forming an air-space of about 2 inches, is another lining of tin, made of sheets seamed together. Both these jackets are bolted to the "panel" front on each side, and are stiffened by the usual cast-iron rings at the top and bottom and in the middle. The top of this cas-ing is made of galvanized iron with a tin lining, and is in
Hot-Air Pipes.   shape a very flat cone. The hot-air pipes are connected to this by means of collars, which are located a little distance from the outer edge. Inside of these collars is a space about 30 inches in diameter. To prevent loss of heat by radiation from this part of the top and at the same time to deflect the air from the center outward an inverted cone of galvanized iron is fastened to the under side of the top, covering this space inside of the collars and forming an air-chamber. A round elbow is fitted to each of these collars of proper degree to give each pipe its due direction and pitch.

This method of connecting the pipes insures a more uni-form distribution of heat than any form of flat top with side connections.

Advantages of Portable Furnace.   The advantages of this "portable" style of setting up fur-naces—that is, those set up with metal casings—are that they take up less room, are more cleanly and economical and are more easily repaired than the "brick-set," and that the cases can be made perfectly tight and are impervious to the air, and do not crack, as is apt to be the case with brick casings.

Combustion.   In this furnace the gases of the combustion from the fire rise into the dome, where they are thoroughly mixed with the air admitted through openings in the feed-door, and are completely consumed, the carbonic oxide being converted to carbonic acid. They then pass through the arms of the dome into the lower chamber of the radiator, and thence by the opening in the diaphragm into the upper chamber and around to the smoke-pipe and chimney. Thus after being completely combined in the dome they are passed twice around the radiator, thereby reducing their temperature before they are allowed to escape into the chimney to as low a degree as is found to be economical.

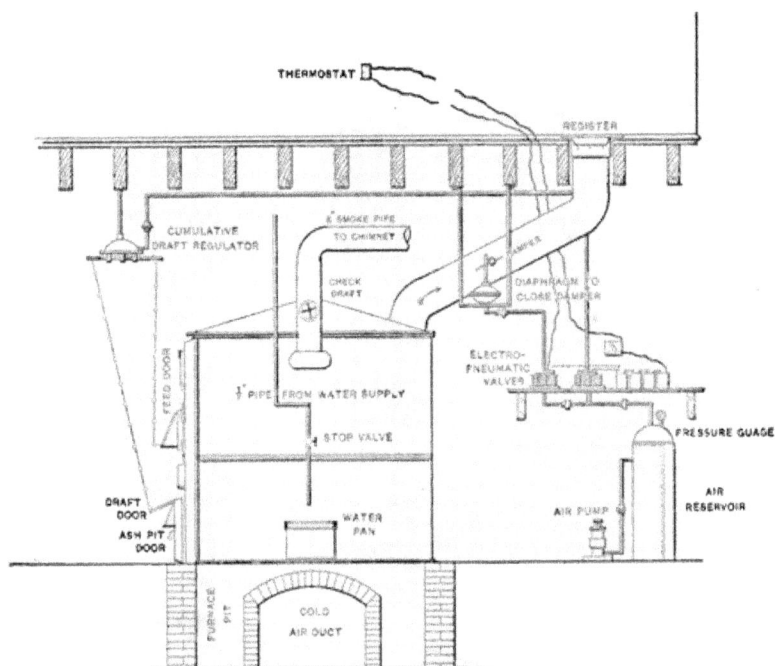

Fig. 5.—Method of Attaching Automatic Regulator to Furnace and Pipes.

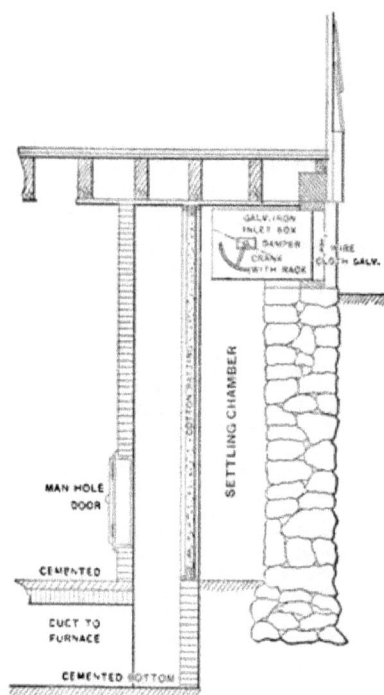

Fig. 6.—Settling-Chamber and Inlet-Box.

These arrangements for securing perfect combustion are of great advantage, first, because they render available the greatest possible amount of heat in the fuel consumed, and second, because they guard against the formation of noxious gases, which might escape into the house.

It is also important that arrangements should be made to insure a fair efficiency under favorable conditions, because the attendants of house-heaters are frequently unskilled, as well as careless and neglectful.

With this furnace, however, the combustion is fair under the most unfavorable conditions, and very little carbonic oxide is formed even with the drafts closed and the check-draft open.

As before stated, this furnace is made almost entirely of cast-iron, which is preferable to wrought-iron or steel as a material for furnaces for the following reasons *Advantages of Cast-Iron.*

It is less attacked by the gases of combustion, does not oxidize as readily, and is in consequence less perishable than either wrought-iron or steel.

It can easily be made of that form which is most favorable for economical combustion and will give the greatest amount of radiating-surface.

With a given amount of radiating-surface it can be made with fewer joints than a wrought-iron or steel furnace, and can be made self-cleaning as far as is possible—that is, without ledges or flat surfaces for ashes to accumulate on. The joints, in this one at least, are made to compensate for the expansion and contraction, thus insuring tightness at all times, whereas with wrought-iron or steel-plate furnaces, the strains caused by contraction and expansion soon open the joints or cause cracks which there are no means of closing.

The cast-iron is less injured by rust during the months when no fires are required, and can be shipped in sections which are easily handled, and can be taken through an ordinary doorway.

It is maintained by many that the injurious gases of combustion will pass through cast-iron, especially when it is heated to redness. As there can never be any outward pressure of gas in a furnace, the special danger on this account is so extremely small as to be neglected, and no furnace large enough for the work required of it ever need be heated to a degree approaching redness.

The importance of introducing into the house a large volume of moderately-heated air has already been mentioned as one of the reasons for selecting a furnace of such large capacity

With a large furnace it is not necessary to heat the radiating parts to a high temperature, which would cause warping or open cracks or burn the grates or fire-pot.

Large Fire. By avoiding this the life of the furnace is not only very much prolonged, but much better results are obtained, because the required amount of heat will be given off from the large body of coals in the fire-pot with a very slow fire.

A small live fire requires frequent attention and fluctuates continually, whereas a large fire will throw off a moderate amount of heat with great uniformity for a long time, and need be attended only at long intervals.

Settling-Chamber In connection with a system of ventilation some means should be provided for removing the dust from the incoming air. In this design it is proposed to introduce the air into a large brick settling-chamber, Fig. 6, through a galvanized-iron inlet-duct, which is fitted with a tight-closing damper, having a crank and rack in a convenient place to regulate the supply of air. The outer end of the duct is covered with a galvanized wire-cloth, about ½-inch mesh, to keep out large articles which might find their way into the house. One side of this settling-chamber is formed of a galvanized-iron frame, across which are stretched two sheets of *Air-Filter.* galvanized wire-cloth ¼-inch mesh, with a space of about ¼ inch between them. This frame is hinged at the back to open like a book, and the space between the two pieces of wire-cloth is loosely filled with thin pieces of cotton-batting.

The area of this filter is 34 square feet, and it is arranged so that it can be slid out for cleaning and removing the cotton, which must be done when it becomes foul and clogged.

Cold-Air Duct. The air passes from the settling-chamber through the filter and thence through the duct to the furnace pit, as shown in Figs. 1 and 6. This air-duct is made with brick sides and arched top and a cemented bottom. This top is cemented over on a level with the cellar bottom, which is thereby free from cracks or joints through which dirt might get into the furnace, and it is on this account superior to a duct with stone or iron covering-plates.

A man-hole door and frame are built into the settling-chamber in case it should be necessary to enter the duct, the furnace pit or the settling-chamber.

Moisture in Air. Air is capable of holding a certain quantity of vapor of water in solution, so to speak, the proportion depending on the temperature of the air; the warmer it is the more it will contain. And when it cools again it deposits it, or forms clouds or fog, which condense on a cold surface. The quantity of vapor per cubic foot of air increases very rapidly [as the temperature advances, a common difference of about 25° in the rise in temperature of the air doubling the quantity of moisture it is able to take up.

During the spring months, in this climate, the percentage of moisture in the air, relative humidity, is about 70. Suppose air at 32°, with a humidity of 70, be taken into a room and warmed to 72°; its humidity will be reduced about one-half below the normal unless some means be taken to supply moisture to this heated air.

In public buildings attempts are sometimes made to secure this amount of moisture by passing the air through sprays or over sheets of water, devices impracticable for a private residence.

The precise influence which either the absolute or relative amount of moisture has upon health is uncertain, for habit enables men to undergo great variations in this respect without ill effect. The disagreeable effects, such as fullness, a tension of the head, flushing of the countenance, cold extremities, &c., which are noticed by the occupants of some apartments, and are attributed to the lack of moisture, are due to a combination of other circumstances, of which the most important is the want of sufficient fresh air to insure satisfactory ventilation, and to the contamination of the heated air by gases from imperfect furnaces.

It is not probable that where a large volume of moderately-heated air is introduced into the rooms, as is proposed in this case, that this difference in the humidity would be sufficiently marked to cause discomfort. This *Evaporating-Pan.* furnace is, however, provided with a large evaporating-pan which is fitted into the casing and which will evaporate several quarts of water in 24 hours. If it is desired to increase the amount of moisture considerably, this can best be done by fastening a large annular evaporating-pan on top of the radiator. This will evaporate as much moisture as the air will hold without allowing it to condense on the cold surfaces, as the windows. Both pans should be supplied with water from a pipe connected to the water-service, and in case the large pan is used it should have a glass gauge to indicate the hight of water in it.

The smoke-pipe to this furnace is 8 inches in diameter and *Smoke-Pipe.* made of No. 20 galvanized iron, with round elbows. It is fitted with a damper which is arranged so that when it is closed more or less a cluck-draft is correspondingly opened.

The flue of the chimney into which this smoke-pipe dis- *Chimney-Flue.* charges is lined with smooth pottery linings, 8 inches square inside. This flue extends below the point where the smoke-pipe enters it, and there access can be had to it for removing dirt through a cast-iron door with frame built into the brick-work.

Air Currents. . The air of a heated room is cooled by the exposed sides and the windows, which causes a downward draft at such points. Thus a circulation is established, the current passing along the ceiling, down on the exposed sides and in front of the windows, along the floor and up on the warm side of the room.

This current of cool air thus set up along the floor is the cause of complaints of cold feet by the occupants of many apartments. These currents can to a large extent be neu-tralized by introducing the heated air or placing the heating-surfaces on the exposed side of the apartment, and when practicable this should always be done.

Hot-Air Pipes. But to do this in a house heated by a furnace it would be necessary to use long and nearly horizontal hot-air pipes. One of the greatest difficulties to be overcome in heating with a furnace is to so arrange and proportion the pipes that each room will receive its due share of the heated air under any and all conditions of the external atmosphere. This would be comparatively simple if these conditions were always the same, but the direction and velocity of the wind are constantly changing, while the proportions and position of the parts of the heating apparatus are not changed to cor-respond with these conditions.

First-Floor Pipes. The pipes to the first-floor rooms could be carried to the outside and arranged to be about of uniform length and re-sistance, but it would be found that the air would not flow through such pipes into rooms on that side of the house against which the wind was blowing. To secure so far as possible this proportionate distribution of the heated air the air-pipes, especially to the first-floor rooms, which are always the most difficult to heat, should be made as short and direct as possible, with an increase in size of the pipes which sup-ply the rooms on the side of the house exposed to the pre-vailing winds.

It should be noted in this connection that the settling-chamber and screen before described act as an equalizer, and prevent sudden gusts of wind forcing the air through the furnace without becoming warmed.

Second and Third Floor Pipes. The pipes supplying the rooms on the second and third floors are distributed and proportioned similarly to those on the first floor, but as a rule much less difficulty is experienced in equalizing the distribution of air on the second than on

the first floor. One pipe was deemed large enough to sup- *Attic Pipe.*
ply the three rooms on the attic floor, as it is not probable
that it would be required to heat all these rooms at once,
and even if it were this pipe would probably be of sufficient
capacity.

The pipe which supplies the lavatory on the first floor is *Lavatory.*
proportionately larger than the others. This is because it is
of necessity longer than the others to the first floor, and also
because it has to discharge air toward the north, under
which circumstances pipes are most apt to prove failures.

The pipe to the bath-room is also made proportionately *Bath-Room.*
larger than the others, because it is frequently desirable to
raise the temperature of this room to a degree much higher
than that required for the other rooms, and it is desirable to
supply this room with a large amount of air to insure thorough
ventilation.

It is not possible to lay down a rule for determining the *Size of*
size of hot-air pipes required to heat a given room. The con- *Pipes.*
ditions of exposure, length and elevation of the pipe, provision for carrying off the foul air, class of building, tightness and number of windows, &c., are so variable that no
definite rule can be laid down.

In buildings of the description of the one under consideration for first-floor rooms a pipe with an area of cross-section
of 1 square inch to each 25 cubic feet of space in the room
to be heated will generally be found sufficient, but for north
rooms this should be increased to 1 square inch for 20 cubic
feet of space. Pipes one-half this size will generally do for
rooms on the second and third floors.

It will be noticed that in this design, where flues are run
up in partitions, the areas of the round pipes from the furnace are somewhat greater than those of the flat vertical
pipes in the partition. This is because the heated air moves
much more freely through vertical pipes than through those
with only a slight elevation.

These hot-air pipes are made of bright tin, the flat pipes
double-seamed together in the usual manner, the round collars double-seamed into the bottoms of the register-boxes.
Wherever these flues are run up in the partitions in a position to come in contact with the lath the latter should be of
iron, and at all other places where the pipes might come in
contact with the wood-work it should be carefully protected
with asbestos sheathing. It is assumed that the flues as
located in this design would be laid out on the plans for the
carpenter, who would so dispose the timbers that the flues
could be gotten in without cutting the frame-work enough
to do serious damage.

Floor Registers. The registers in the parlor, hall and library are placed in the floor. Floor registers have the disadvantage that they collect more or less dirt and are a source of loss of small articles, but, as has been stated, the importance of securing a proportionate distribution of the heated air is paramount, and the absence of sharp turns and changes in the form of a section of pipes as arranged to connect floor registers is found to greatly facilitate this.

Side-Wall Registers. The other registers are set in the side walls where it is more convenient to put them and where they would answer every probable requirement.

Pipe-Collars. It is best to put a deep collar in pipes which supply floor registers, at the heel of the elbow which turns into the box. This is fitted with a removable cap and serves as a receptacle to catch the dirt or anything that might fall into the pipe, whence it can be easily recovered.

Register-Boxes. Around these floor-register boxes there should be fitted a square tin box somewhat deeper than the timbers. This just fits into the opening cut in the floor for the register border, where it is fastened, and is flanged off on the under side of the cellar ceiling. This makes a finish and covers up the ragged edge of the plaster.

Lavatory Ventilation. The register which supplies air to the lavatory on the first floor is set in the side wall 6 feet above the floor, because there is no other place for it except in the floor, which is especially objectionable in a bath or toilet room.

This method of introducing the warmed air near the ceiling and taking the cold out at the floor is very desirable, because it insures a uniformity of temperature in all parts of the room and immunity from unpleasant currents, but is generally applicable only with those forms of apparatus which have mechanical appliances for moving the air.

In each of the toilet-rooms there is a ventilating-register which is connected with a separate flue in the kitchen chimney. As this chimney is always warm, there will at all times be a draft of air through these ventilating-flues and a tendency to draw the air from the other rooms into these two, preventing the dissemination of unpleasant odors from them.

Register-Faces. It is recommended to use on the first floor registers with solid bronze or brass faces and borders. The first cost of these is somewhat greater than that of other kinds, but they preserve their good appearance, even with a great amount of wear, almost indefinitely. The same kind should be used for the principal rooms on the second floor, but in other places well plated or japanned registers would answer.

The areas of the openings of registers should be at least 25 per cent. larger than those of the pipes to which they are attached, on account of the friction they produce, and also to reduce the velocity of the incoming air, by allowing it to expand.

Area of Registers.

Excepting in the two toilet-rooms, in this design provision is made for drawing out the depreciated air throughout the open fire-places, of which there are three on each floor. The flues with which these fire-places are connected have an area of section of about 64 square inches each. This is not large enough to carry off the estimated supply of air, but larger flues could not be built into the chimney without making radical changes in their dimensions. These flues, as well as the flues for the ceiling ventilators, should have a smooth lining of pottery.

Ventilating-Flues.

Automatic Draft-Regulator.

It has already been stated that perfect ventilation may be said to have been secured in an inhabited room only when it has been established in connection with the perfect comfort of the occupants as regards temperature, being neither too hot nor too cold.

With such an apparatus as has been described these results would not be obtained unless the fire be controlled by other means than the ordinary drafts and dampers operated by hand, even in connection with the convenient devices in use for operating these dampers from the "living" room of the house.

If the attendant to the furnace were usually vigilant very fair results may be had at some times, particularly in cold weather, as the tendency is always to overheat in mild weather.

The occupants of an apartment do not notice that the furnace requires attention until the room becomes uncomfortably cold; then they open the drafts and the fire starts up. After some little time the temperature is raised to a comfortable point, but they do not think to check the fire then, and this continues to burn until the room becomes too hot or *vice versa*, and the consequence is that the temperature is generally considerably either above or below the desired degree and fluctuating continually.

This has led to the invention of many automatic draft-regulators, some of which have been perfected to such a point that if properly applied they will maintain a temperature within one or two degrees of the standard, unless, of course, the fire be allowed to go out.

The apparatus which is proposed in this design will be readily understood from the detail drawing, Fig. 5.

The peculiar feature of this system is the electro-pneumatic valve—that is, a pneumatic-valve operated by electricity. The motive power for working the dampers is compressed air, which is stored in an iron tank. This tank is provided with a pressure-gauge and a small hand-pump, with which the attendant keeps the air-pressure in the reservoir to the required point. This requires only a minute or two daily, and if desired an automatic pump, operated by water under pressure from the service, can be used instead.

In each room of which it is desired to control the temperature is placed a thermostat, which is connected by electric wires to a corresponding electro-pneumatic valve.

The hot-air pipe to each room is fitted with a butterfly damper, to the rod of which is attached a small pinion. Fastened to each air-pipe is a small diaphragm apparatus, which moves a rack, which engages the pinion and turns the damper. Each diaphragm is connected to its respective electro-pneumatic valve with a small composition tube and each valve is connected to the air reservoir in the same manner. Suitable electric batteries are provided in the cellar for furnishing the requisite current of electricity.

The action of these appliances is as follows: When the temperature of the room rises to a certain point an electric contact is made; this causes the electro-pneumatic valve to open and admit the compressed air to the corresponding diaphragm, and this, in turn, closes the damper in the air-pipe. When the temperature falls this contact is broken, the pneumatic valve shuts off the air from the reservoir and allows the compressed air in the diaphragm to escape, which causes the damper to open.

This apparatus in a large measure overcomes the defect in all hot-air furnaces arising from the fact that it is next to impossible to so place the furnace and arrange and proportion the flues as to equally heat all the rooms of a house, especially if the arrangement of the house is not compact.

The pipes leading to the different rooms have different lengths and offer different resistances to the air, the short pipes and the vertical pipes giving the greatest amount of hot air, which tends to overheat the rooms they supply and to rob those rooms having longer pipes of the heat they should have.

The direction of the wind also causes the heating apparatus to vary from day to day, as the heated air tends to go to the side of the house opposite that against which the wind blows.

With this automatic regulating apparatus the hot air is shut off from the rooms which are warm enough, and this obliges the air to go to the other rooms, to which it would not otherwise go. Therefore the rooms that are the most easily warmed will be the soonest shut off, and the heated air which would tend to overheat these rooms is sent to rooms that do not heat so readily, such as are further away from the furnace or have smaller flues in proportion, or are toward that side of the house against which the wind blows.

Shutting off the heat from each room, however, will not control the fires, and some provision must be made for this, otherwise the furnace might become overheated. An attachment called a cumulative draft-regulator is provided for thus purpose. This consists of a diaphragm apparatus which opens and closes the dampers below the grate and the check-damper. This diaphragm is operated by air from the reservoir admitted through another electro-pneumatic valve, which is controlled by all the thermostats in the house. Thus, when all the rooms of the house are warm

Room.	Hight.	Cubic contents in feet.	Area of round air-pipe in cellar. Square inches.	Area of air-pipe in partition. Square inches.	Size of admission register.	Size of ventilating register.	Size of heat flue. Square inches.	Area of fire-place flue. Square inches.
Parlor..............	10¼	3,045	154	....	14 x 18	12 round	28	64
Library.............	10½	2,835	113	....	12 x 16	12 round	28	64
Lower hall..........	10⅓	2,625	113	....	12 x 16	....	..	..
Upper hall..........	9½	2,424	...	...	..	..	..	..
Dining-room.........	10½	2,919	113	42	12 x 16	12 round	28	64
Toilet..............	10½	189	63	31½	9 x 12	9 x 12	31½	
Second floor:								
North chamber	9½	2,327	63	42	10 x 14	....	..	64
Front chamber.......	9¼	2,125	63	42	9 x 12	....	..	64
West chamber.......	9½	2,641	63	42	9 x 12	....	..	64
Bath...............	9½	427	63	31	9 x 12	....	..	..
Rear chamber........	8¼	1,666	78	42	9 x 12	9 x 12	31½	..
Sewing-room	8¼	571	78	42	9 x 12	....	..	
Third floor:								
Play-room...........	9	2,394	63	42	9 x 12	....	..	
Billiard-room........	9	1,629	63	42	9 x 12	....	..	
Chamber............	9	1,489	63	42	9 x 12	....	..	
Totals.............	....	32,232	949	....	..	..	....	

Cubic contents: Total, 32,232 feet. **First floor, 14,037.** Second floor, 12,687. Third floor, 5508.

Area of air-pipes. **Total, 949 square inches.**

Area of fresh-air inlet, 600.

Area of air-pipes: First floor, 556 square inches.

Cubic feet of air warmed by 1 square inch area of air-pipes: First floor, 27. Second floor, 47½. Third floor, 87.

Area of grate surface furnace, 803 square inches.

Cubic feet of air warmed by 1 square inch of grate surface, 40.

Table Giving Dimensions of the Principal Parts of Apparatus.

and the heat shut off from them by their respective thermostats, this electro-pneumatic valve will admit the air to the diapragm apparatus, which will close the damper in the ash-pit door and open the check-draft.

The economy of this arrangement is apparent.

It may be stated in this connection that the best way to check the fire is to open the door and admit air over the fire, as shown in the detail drawing.

There are many details of this apparatus which it is not necessary to describe here.

Double thermostats are provided for the living-rooms, whereby a lower temperature may be kept at night if so desired.

In the table given above, where two or more rooms are supplied from one pipe, the full size of the pipe is entered for each room, but allowance is made for this in footing the totals.

ESTIMATE.

Furnace (with cases).. ...		$220.00
Registers with bronze faces and borders.—1 14 x 18 ; 2 12 x 16		
Registers only, bronze faces—1 12 x 16; 1 10 x 14; 6 9 x 12............		100.00
Nickel-plated registers—1 9 x 12 with frame; 4 9 x 12..................		23.50
3 12-inch ceiling ventilators........		15.00
Register-boxes—1 14 x 18; 3 12 x 16, @ 75 cents each...............		3.00
1 10 x 14; 11 9 x 12, @ 50 cents each............................		6.00
3 boxes for finishing around register openings in cellar............... ..		2.25
12 collars, @ 25 cents each....................................		3.00
5 feet 14-inch pipe ; 25 feet 12-inch pipe, @ 30 cents per foot...........		9.00
12 feet 8-inch galvanized smoke-pipe, 4 pounds per foot, @ 7 cents per pound..		3.36
68 feet 3½ 12 pipe, 55 feet 3¼ 9 pipe, 28 feet 3¼ 8 pipe, @ 25 cents per foot		37.75
Furnace pit.—Digging................................	$4.00	
1 1-10 M brick @ $18............................	19.80	
	——	
		23.80
Air duct.—Digging................................	4.00	
9-10 M brick, @ $18..................................	16.20	
	——	
		20 20
Settling chamber.—1 2-10 M brick, @ $18..................		21.60
Inlet box.—75 pounds galvanized iron, @ 10 cents per pound...	7.50	
Damper......	3.00	
	——	
		10.50
Screen and frame..		15.00
15 days' labor, man and helper, @ $5............................		75.00
		————
Total..		$588.96
Automatic draft regulator (complete)...................................		295.00
		————
Total..		$883.96

(In this estimate no allowances are made for freights, cartages or cutting for the pipes.)

HOT-AIR SYSTEM.*

BY JAMES A. HARDING.

HEATING.

Upon the plans submitted by the writer is shown the installation of a heating plant for warming the house by means of a brick-set hot-air furnace.

The object has been to so arrange the apparatus as to furnish the maximum amount of warm air to the rooms of the house with the minimum consumption of fuel, the consideration of first cost having been secondary to that of economy in running the apparatus.

Certain pipes which might be regarded as unnecessary have been, in this plan, made to subserve the end of convenience and provide for a correctly-proportioned distribution of warm air in all of the three stories of the house.

Warm-Air Distribution.

As framing plans of the building are not submitted for consideration, it is assumed that the building will be framed to accommodate any position of pipes not seriously conflicting with strength of construction, and it is believed the plans under consideration do not.

The writer has kept in view the statement in your specifications that "competitors will be allowed to make suggestions as to flues, location of apparatus, provision for pipes, &c."

It has been the object to so locate the furnace as to provide for the shortest and most direct runs of cellar hot-air pipe, and at the same time to so place it as not to obstruct convenient passage-ways in the cellar. Having provided a 3-foot 6-inch opening between the furnace and the stair partition, it is believed that such obstruction will not be thought to exist.

Location of Furnace.

By having the front and feed door of the furnace at the door-way, which opens into cellar A, the dust incident to the removal of ashes from the furnace may be confined to that compartment of the cellar.

Prevention of Dust.

It is desirable that the heat radiated from the cellar pipes and the furnace be applied as a means of warming the floors of the living-rooms, more particularly the library and the

* From *The Metal Worker*, September 21, 1889. Copyrighted, 1889, by David Williams.

dining-room. This result is accomplished by the free circulation of air so warmed throughout those portions of the cellar under the main hall and cellars A and B.

Temperature
of Cellar. In locating the furnace as shown, the contributor has borne in mind that "the temperature of the cellar is not regarded as of special consequence, but that some portion of it must be reserved for storage of fruits and vegetables." It will be observed upon inspection of the cellar plan that there are no heating-fixtures of any kind in either of the cellars marked A, B or C, and that but one pipe enters the cellar-room D. It is thought that for convenience of access from the kitchen either the cellar-room C or D, or both, might be used for cold storage, and in order that the cellar D may not be heated in any degree, the hot-air pipe entering it be wrapped with an asbestos preparation. This wrapping will also protect the said pipe from the cooling influences of that part of the cellar. For the latter reason, it is advised that the hot-air pipes in the hall at foot of cellar stairs also be wrapped in a similar manner.

Furnace
Setting. It is proposed to have the furnace set in brick-work for several reasons : 1. The double brick walls inclosing it prevent a large proportion of the radiation of heat which takes place in a portable furnace. 2. The front of the brick-work being built tightly against the foundation wall of the house confines the dust to the compartment A, as before stated. 3. The brick structure when once thoroughly heated becomes itself a reservoir of heat, which is given off gradually after the fire in the furnace may have been allowed to burn low through neglect ; whereas in a portable furnace no such storage of heat is possible. 4. Lime being a preservative of iron, the brick structure will protect the furnace from rust when it is not in use.

A further advantage of a brick-set furnace is in the convenience of getting at the furnace to clean or repair it when it becomes necessary.

Cold-Air
Supply. The cold-air supply to the furnace is provided for by a duct under the concrete floor of cellar C. A vertical shaft of brick built up from this duct at the window is to be so constructed that the window-sash may be set in its face or front wall to give light, as if set in the usual manner, and to be opened when furnace is not in use for the purpose of ventilating the cellar (see Fig. 1). The damper for regulating the supply of cold air is to slide horizontally below the level of the window-opening (see Fig. 1).

It is desirable to introduce the cold air at the back of a furnace, for the reason that as the air enters the chamber it is divided by the ash-pit of the furnace and diverted so as to supply hot-air pipes leading from both sides of the furnace. The proper size cold-air box, as related to the total capacity of hot-air pipes leading from the furnace, varies under different conditions of wind and atmosphere. With a high wind and a rare and dry atmosphere less supply is necessary than with a still, heavy or damp atmosphere.

In placing the registers in the different rooms, it has been the object to locate them at the side of the room opposite the fire-place or ventilator, for the reason that the exhaust of air by process of ventilation induces the warm air to circulate in the direction of such fire-place or ventilator and affords thereby a more complete and perfect warming of the rooms. *Position of Registers.*

The occupation by registers of wall spaces suitable or desirable for placing of furniture has been avoided. The registers in the first-story rooms have been set in the side wall, with the idea that many people object to cutting carpets; also that when placed in the floors they are likely to become receptacles for the accumulation of dirt.

Registers in each of the principal rooms of the house have independent hot-air pipes from the furnace to connect with them, the exceptions to this rule being the three small registers in the second-story-hall, the one in the sewing-room and the one in rear hall, first story. The amount of heat necessary to be obtained from these five registers is inconsiderable. *Hot-Air Pipes.*

The writer is aware that where a hot-air pipe connects with two registers placed at different altitudes the higher one of the two is liable to receive nearly all the heat, particularly when the difference in hight is great. Where this plan has been adopted in the present case, a judicious reduction in size of hot-air pipe just above the lower register has been provided for in order to make the proper distribution of warm air. This plan has not been adopted in any case where much heat is required from the lower register.

It has been thought well to connect a register with the warming-closet in the butler's pantry, for heating the kitchen at such times as the fire in the range might be out, in order to protect the plumbing from frost. *Warming-Closet.*

VENTILATION. ·

An elaborate system of ventilation has not been thought necessary, but an effective plan for securing perfect circulation of warm air and sanitary condition of rooms which are plumbed is given. It is conceived that such rooms as have fire-places with good flues will require no other means of ventilation.

<u>Kitchen Ventilation.</u>   Defective ventilation of the kitchen is most noticeable in its failure to remove the odors of cooking. To provide for this and the ventilation of several other rooms as well it is proposed to construct the chimney with a flue 8 x 24 inches, having a tile smoke-pipe for the laundry stove and range 8 inches in diameter built in the center, leaving a space of ventilating-flue on either side of it 8 x 8 inches. The ventilator in the kitchen should be placed near the ceiling, and open into one of the ventilating-flues at the side of the tile smoke-pipe, and a hood should be placed over the range for the concentration and direction of fumes of cooking into the flue.

<u>Back Hall.</u>   The large deposit of cold air in the back hall, which will naturally descend the back staircase and enter from the porch door, is to be removed by means of a ventilator in the floor, connected, by a pipe passing through the laundry, with the ventilating-flue in the kitchen chimney on the opposite side of smoke-flue from the kitchen ventilator.

<u>Kitchen Chamber.</u>   The ventilation of the chamber over the kitchen to be provided for by a ventilator set in the chimney and opening into this ventilating-flue.

<u>Sewing-Room and Bath-Room.</u>   The ventilation of the sewing-room and bath-room to be provided for by means of a tin pipe passing up through the partition into the attic, and connecting with one of the ventilating-flues in the kitchen chimney, just below where it passes through the roof.

<u>Toilet-Room.</u>   The ventilation of the toilet-room, first story, to be accomplished by a pipe connected with a ventilating-flue in parlor chimney at the side of a tile smoke-pipe, which is to be built into this chimney for a furnace smoke-flue. The ventilation of the wash-basin in chamber over parlor is provided for by a ventilator opening directly into the ventilating-flue adjacent to furnace smoke-flue.

<u>Play-Room.</u>   Ventilation of childrens' play-room by means of a ventilator set in chimney and opening into ventilating-flue adjacent to furnace smoke-flue.

<u>Wash-Basins.</u>   Ventilation of wash-basins in chambers over dining-room and library respectively is provided for by means of a pipe leading up through partition into space under rafters and connecting with an independent ventilating-flue in the dining-room chimney.

<u>Third-Story Bed-Room.</u>   Ventilation of the third-story chamber by means of a pipe connecting with this ventilating-flue in dining-room chimney.

<u>Billiard-Room.</u>   Ventilation of billiard-room by means of two ventilators; one at floor and one at ceiling, and opening into an independent ventilating-flue in library chimney.

The ventilators arranged in the manner described will, in connection with the natural ventilation through the fireplaces, provide for an effective exhaust of impure air from all of the rooms of the house and induce a constant change of atmosphere.

All ventilators should be set at the floor-level and be fast- Location of Ventilators. ened to the base-boards of the rooms by means of screws; one ventilator to be placed near the ceiling in the billiard-room for the exit of smoke.

The ventilators are placed near the floor for the reason that carbonic oxide and other deleterious impurities of the air, being heavier than the atmosphere, will settle to the floor-level of the rooms. A further object is to remove the colder stratum of air from the rooms and thereby promote uniformity of temperature and economize the warmed air from the furnace.

AUTOMATIC REGULATION.

In respect to many of the so-called devices in the market this term is a misnomer.

For the regulation of hot-air furnaces there are several Expanding Rod Regulator. devices which have for their salient feature an expansion-tube containing an iron or steel rod placed directly above the furnace proper within the hot-air chamber. The contraction of the more sensitive tube with the cooling of the furnace causes the iron rod to protrude from the end of the tube and by working a lever to open the drafts of the furnace.

The increase of heat incident to the opening of drafts causes the tube to expand and lessen the protuberance of the iron rod, so as to gradually close the drafts of the furnace and check the fire.

This process is alternately repeated, but so gradually that the fire is kept burning at a very even rate.

The apparatus is provided with a figured dial, by which it may be set—that is, the relative length of rod and tube changed—to increase or diminish the amount of fire, according to changes in the weather.

While this form of regulator is admirable as a means of securing uniform combustion and economy, it is manifestly imperfect as a regulator of the temperature in the living-rooms, in that it bears no necessary relation to their temperatures and is not affected or operated by such temperatures.

Another form of regulator is an electrical contrivance op- Electrical Regulator. erated by means of a thermostat placed in one of the living-rooms. The thermostat is sensitive to a rise or fall of less than $1°$ of temperature, and will, at this slight change, close an electric circuit, which, by means of various devices more or less effective, will open and close the drafts of the furnace to increase or reduce the heat as may be necessary to maintain the temperature at which the thermostat is set.

This form of regulator (of good make) will maintain an even temperature in the room where it is placed.

If it is placed in the first story and the doors of the various rooms are left open, so that the temperature of the entire floor is very much the same, then the apparatus will be found most effective and the result of its use satisfactory.

Compressed-Air Regulator.   Another form of regulator is one which has a compressed-air chamber in the cellar, with pipes leading to valves for operating the dampers in the various hot-air pipes leading from the furnace, as well as the drafts of the furnace. A thermostat is placed in each of the rooms heated by the furnace, and each thermostat is arranged to operate (by means of a jet of compressed air let into the valve referred to) a damper in the hot-air pipe leading to such room.

By this means the dampers in all hot-air pipes are successively closed when the desired temperature in the rooms has been reached.

Thermostat.   A thermostat is also placed in the room which is found by experiment to receive its quota of heat last, and arranged to close the drafts of the furnace when all the other rooms are sufficiently warmed.

This form of regulator is the most complete and effective of any but complicated and expensive.

Cost of Regulators.   The style of regulator first described will cost $10 to $20; the second from $40 to $50; the last from $150 to $250.

The writer has used one of the second style for two years, and in his house, where the doors are seldom closed, it has been found an economist in fuel, time and attention and a positive blessing.

MECHANICAL CONSTRUCTION.
Furnace.

Assuming the furnace to be of cast-iron, the fire-surface should be at least 32 inches in diameter.

Assuming the radiator of the furnace to be 4 feet 6 inches in diameter, with an adequate space between it and the dome or body of the furnace for the passage of hot air, it is deemed advisable to have the circular or interior brick wall of the furnace built 14 inches larger in diameter than the radiator— that is to say, with a space of 7 inches between it and the radiator all round. This interior wall and the covering form the hot-air chamber. See Figs. 1 and 2.

The best method of securing the contact of cold air with the furnace is to set circular iron plates upon brick piers at

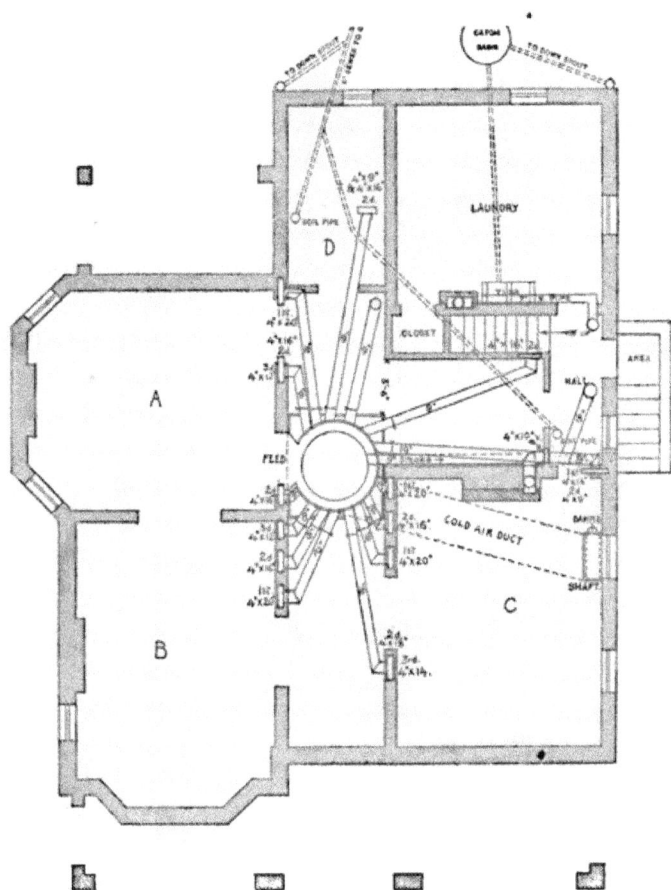

Fig. 1.—Cellar Plan.—Scale, 1-12 Inch to the Foot.

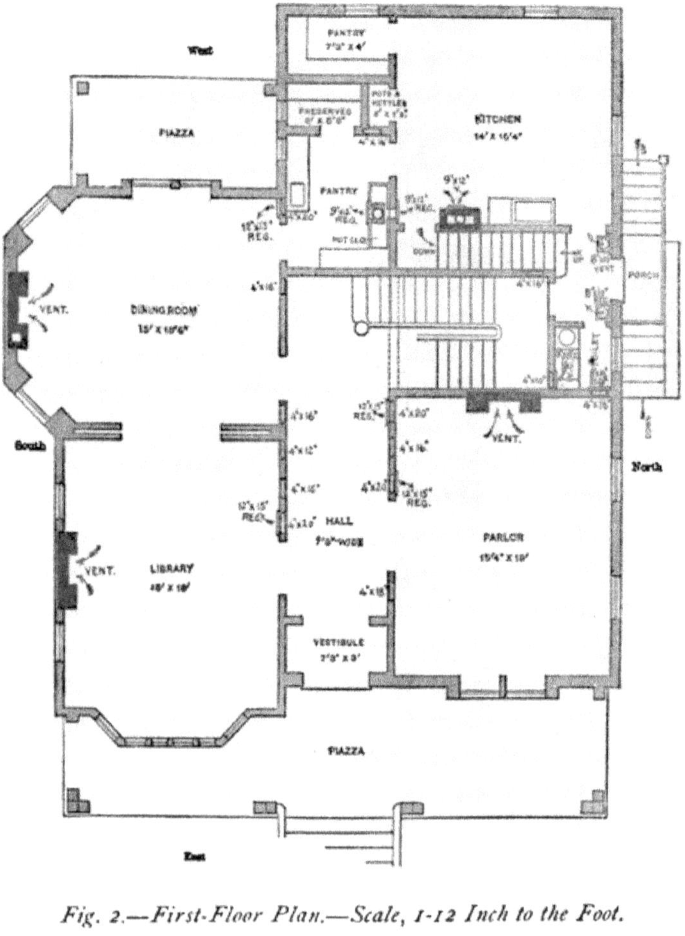

PANTRY
7'2' X 4'

West

PRESERVED
6' X 0'0'

POTS &
KETTLES
4' X 7'6'

PIAZZA

KITCHEN
14' X 16'4'

PANTRY
9'x12'
REG.

9'x12'

12'x15'
REG.

HOT AIR

VENT

PORCH

4'x16'

DOWN

4'x16'

VENT.

DINING ROOM
15' X 18'6'

4'x16'

4'x16'

12'x20'
REG.

4'x20'

4'x16'

4'x16'

VENT.

South

North

4'x16'

4'x12'

4'x16'

4'x16'

12'x15'
REG.

4'x20'

12'x15'
REG.

4'x20'

HALL
7'0' WIDE

PARLOR
15'4' X 19'

VENT.

LIBRARY
18' X 19'

4'x16'

VESTIBULE
7'6' X 9'

PIAZZA

East

Fig. 2.—First-Floor Plan.—Scale, 1-12 Inch to the Foot.

the level of the bottom of the fire-pot ; said plates to project inward toward the base of the furnace and support the circular wall upon their outer edge. The space between the piers which support these plates and between the inner edge of the plates and the base of the furnace to be equal to the capacity of the cold-air box or duct. See Figs. 1 and 2.

The outer square brick wall of the furnace should have a space of 2 inches at least between it and the circular brick wall at the nearest point to form a non-conducting air-chamber. See Figs. 1 and 2. **Furnace-Chamber.**

The hot-air pipes must connect with the interior or hot-air chamber of the brick structure at its highest point, which, in a cellar 8 feet 6 inches, as in the present case, should be at least 18 inches below the ceiling. This will afford a good pitch to the hot-air pipes, and at the same time give ample headroom in the cellar. See Fig. 1. **Hot-Air Pipe Connection.**

The cold-air duct should be at least 16 x 36 inches, to provide for the maximum supply of cold air that will ever be required. An excavation should be made of sufficient depth to admit of an inch of cement in the bottom of the duct itself, 2-inch flag-stone to cover it, and the usual thickness of concrete over that. See sectional view, Fig. 1. **Cold-Air Duct.**

Sizes of all of the horizontal cellar hot-air pipes and vertical partition hot-air pipes will be found indicated on the cellar plan. The proper installation of the partition-pipes will be described in Figs. 3, 4 and 5. The ventilating partition-pipes to be put in similar manner, with the exception that iron lath will be unnecessary. **Size of Pipes.**

The side-wall registers should be set in plaster-of-paris, and any such as are in panel or wainscoting should have soapstone or other fire-proof margins. They should be set just above the base-board and if of oblong shape in a horizontal position, so placed that when the valves are partially opened the air will pass between them in an upward direction. See Figs. 6, 7 and 8. **Side-Wall Registers.**

The object in placing registers horizontally is to furnish hot-air openings as near the full width of the partition-pipes as possible, and to avoid the concentration of warm air to the center for exit.

All registers set in stud partitions should be of the convex pattern. This form of register is but a trifle over 1 inch in depth when the valves are opened, which abuot corresponds to the thickness of plaster in front of the hot-air pipe, so that the register-valves will not project into the pipe and cut off the passage of hot air through the register in question or one above in the same pipe. See Figs. 6 and 7.

Tin Casing. Tin casings for the protection of floor-beams and lathing of the cellar ceiling should be provided for the floor registers. See Fig. 9.

Butler's Pantry. The register for heating the butler's pantry to be placed in the floor of the warming-closet in the usual manner. The heat to circulate through the shelves in the warming-closet and to pass out of register (placed in the door of the closet) into the butler's pantry. The shelves should be constructed of iron slats set sufficiently far apart to support a plate edge-wise, so that the circulation of warm air will be around and in contact with all of the plates to be warmed.

The interior of the closet should be lined with galvanized iron. The lining not to be nailed directly to the closet, but screwed to the same, thick washers being used to keep the iron from the wood-work.

Description of Sketches. Description of Figs. 1 and 2:

Fig. 1. Fig. 1 is a sectional view of the furnace, brick structure inclosing same and cold-air box.

Fig. 2. Fig. 2 is a plan of the setting of the furnace on line A–B, Fig. 1.

A, ash-pit or base of the furnace.

B, space between base and iron plates for passage of cold air.

C, iron plates.

D, brick piers 4 x 8 inches, supporting iron plates and interior wall.

E, inner 4-inch brick wall, plastered smooth on the inside

F, non-conducting air-spaces between inner and outer' walls.

G, outside 4-inch brick wall.

H, two courses of brick resting upon tin-plates, supported by iron bars and forming cover to hot-air chamber.

I, one course of the brick supported by iron bars, covering the whole structure.

J, hot-air chamber.

K, window-opening in outside wall for ingress of cold air.

L, wire screen for the exclusion of vermin.

M, window-sash set in face of brick wall, swung upon hinges.

N, the slot in the brick wall through which the horizontal damper, or cut-off, works.

O, 4-inch brick wall forming a cold-air shaft from the window to duct under the floor.

P, the man-hole door, giving access to duct when necessary.

R, cold-air duct, 16 x 36 inches, under the cellar floor.

S, concrete floor of the cellar.

T, flag-stone covering the duct and supporting the concrete.

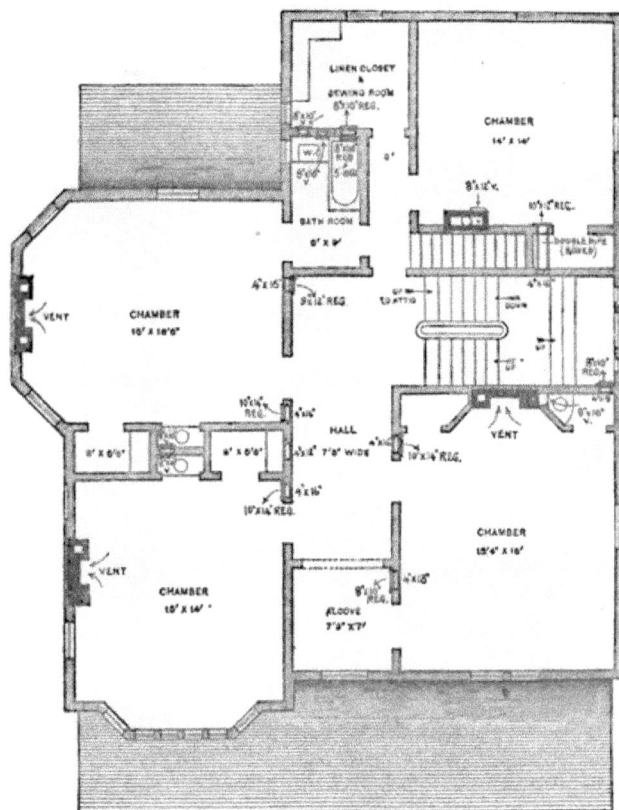

Fig. 3.—Second-Floor Plan.—Scale, 1-12 Inch to the Foot.

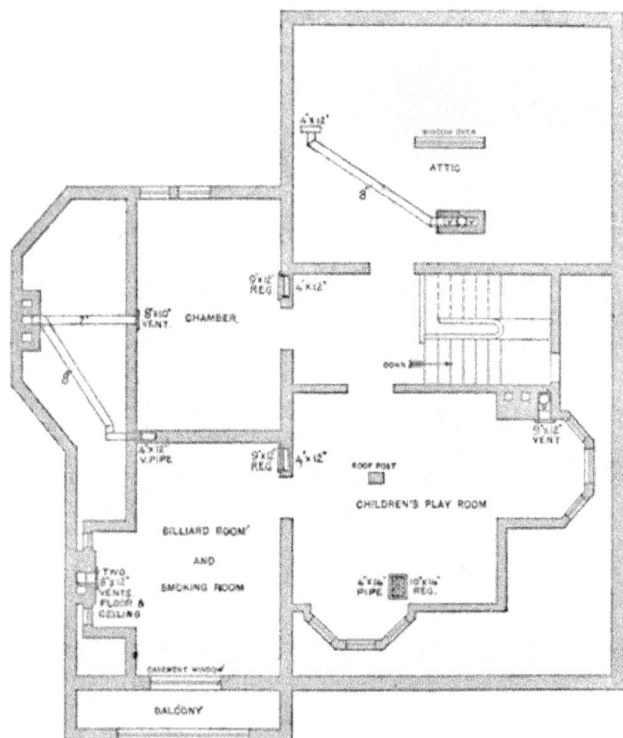

Fig. 4.—Attic Plan.—Scale, 1-12 Inch to the Foot.

Fig. 3 is a horizontal sectional view of partition, showing installation of hot-air pipe. Fig. 3.

Fig. 4 is an elevation of the partition, showing the same thing. Fig. 4.

Fig. 5 is a horizontal sectional view of the partition at floor-level, showing the same thing. Fig. 5.

Description of Figs. 3, 4 and 5 :

A, 2 x 4 inch studs.

B, 1-inch coat of plaster

C, iron laths covering the hot-air pipe in partition, so placed as to break joints with wooden lathing and obviate cracking of walls in case of straight joint.

D, tin lining on the face of the studs adjacent to hot-air pipe for protection of studs ; the same bent to set ½ inch away from stud.

E, hot-air pipe set with a space 1½ inches at each side between it and the studs.

F, wooden laths covering spaces in partition not inclosing hot-air pipe.

G, edge of floor cut back 1 inch from face of hot-air pipe.

H, base-board.

I, plastering on wall between base-board and hot-air pipe.

Note.—The partition hot-air pipes between floors and ceilings below to be double.

Fig. 6 is a vertical section of partition and hot-air pipe showing the proper manner of setting register and kind of register to use. Fig. 6.

Fig. 7 is a horizontal sectional view showing the same thing. Fig. 7.

Fig. 8 is an elevation of partition hot-air pipe, showing register opening. Fig. 8.

Descriptions of Figs. 6, 7 and 8 :

A, 2 x 4 inch studs.

B, 1 inch of plastering on face of studs.

C, base-board.

D, register set directly above base-board, showing position of valves when partially opened.

E, hot-air pipe.

Description of Fig. 9 : Fig. 9.

A, soapstone border set in bed of plaster.

B, register face resting in rabbet of soapstone.

D, hot-air pipe.

E, tin casing protecting floor-joists and cellar ceiling.

F, air-space between hot-air pipe, register-box and tin casing.

G, floor-joists.

H, floor.

COST.

The estimate of cost is based upon what would be considered a fair charge for labor and materials in the vicinity of New York City.

The quality of the several kinds of material employed to be of the best.

IX bright charcoal tin is to be employed for all work except lining studs, for which IC bright is to be used.

In may be necessary to add to the estimate submitted crating of pipes, cartage, freight, traveling expenses of men, board of men, expense of superintendent, &c., none of which items can be accurately estimated in the present case for obvious reasons.

Several of the items included in the estimate are usually furnished by other contractors than the heating man, but in such cases the owner of the building cannot accurately estimate the total cost of his heating apparatus.

It is sometimes an advantage to have the rough material, such as brick and mortar, furnished by other contractors, for the reason that they are likely, for various reasons, to put such materials in at less price.

ESTIMATE OF COST OF HEATING.

Furnace. Brick-set furnace having 32-inch fire-pot, with iron bars and plates, set of fire tools and evaporating-pan.............. $185.00

Cellar Hot-Air Pipes. 51 feet 10-inch tin pipe, 6 elbows, 6 dampers.
58 feet 9-inch tin pipe, 2 elbows, 9 dampers.
7 feet 8-inch tin pipe, 2 elbows, 3 dampers.

Smoke-Pipe. 12 feet 9-inch No. 24 galvanized-iron smoke-pipe.
2 elbows, 1 regulating-damper.

Vertical Hot-Air Pipes. 16 feet 4 x 20 inch tin pipe, 8 heads.
16 feet 4 x 18 inch tin pipe, 2 heads.
100 feet 4 x 16 inch tin pipe, 14 heads.
8 feet 4 x 14 inch tin pipe, 2 collars.
37 feet 4 x 12 inch tin pipe, 3 heads, 1 collar.
15 feet 4 x 9 inch tin pipe, 2 heads, 2 collars.
3 feet 10-inch tin pipe, 2 collars.
3 feet 9-inch tin pipe, 2 collars.
5 10-inch, 8 9-inch, 3 8-inch collars.

Register-Boxes. 4 12 x 15 register-boxes.
4 10 x 14 register-boxes, 1 casing.
1 10 x 12 register-box.
5 9 x 12 register-boxes, 1 casing.
6 8 x 10 register-boxes, 1 casing.

Cylinders between Floors and Ceiling. 4 5 x 21 inch tin cylinders, 14 inches long.
2 5 x 19 inch tin cylinders, 14 inches long.
13 5 x 17 inch tin cylinders, 14 inches long.
3 5 x 13 inch tin cylinders, 14 inches long.
1 5 x 10 inch tin cylinder, 14 inches long.

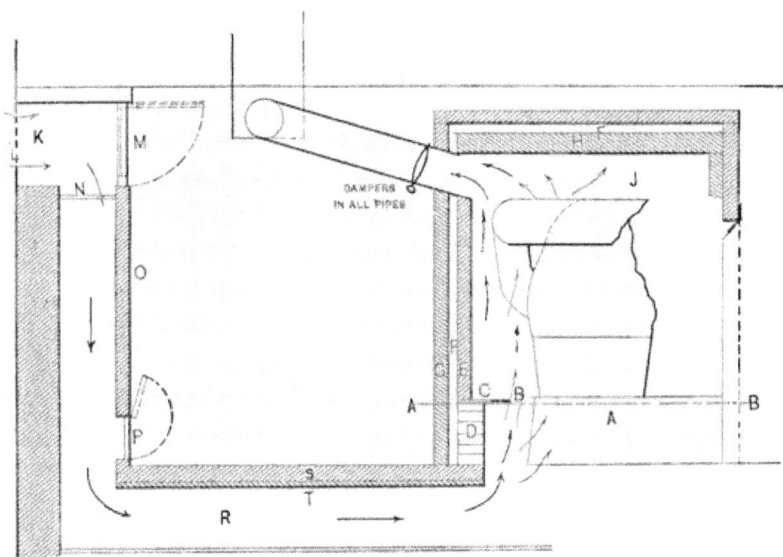

Fig. 1.—Vertical Section through Cold-Air Duct and Furnace Chamber.

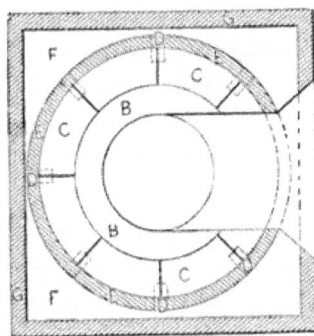

Fig. 2.—Horizontal Section through Furnace Chamber.

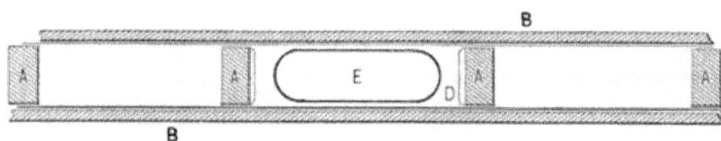

Fig. 3.—*Horizontal Section through Partition-Pipe.*

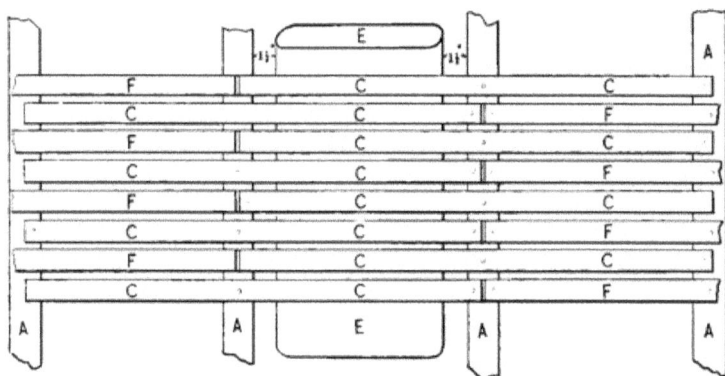

Fig. 4.—*Elevation Showing Partition-Pipe.*

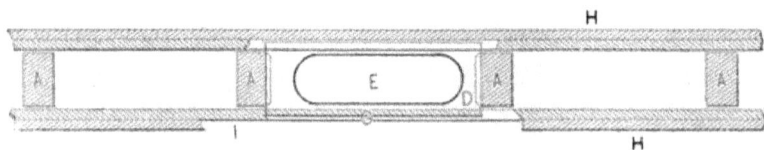

Fig. 5.—*Horizontal Section through Partition-Pipe at Floor-Level.*

600 pounds iron lath. Iron Lath.

65 sheets 14 x 20 inch IC tin cut into pieces of proper width for lining. Stud-Lining.

10 pounds roofing-nails.

1 14-inch sheet-iron thimble with cover for smoke-pipe; to be built into walls of furnace. Smoke-Pipe Thimble.

20 sheets 14 x 20 tin for covering hot-air chamber, cut 8 inches wide.................................. **$143.84** Tin Cover.

1 10 x 14 square, black or white japanned. Registers.
3 9 x 12 square, black or white japanned.
1 8 x 10 square, black or white japanned.
4 12 x 15 square, black or white japanned, convex.
3 10 x 14 square, black or white japanned, convex.
2 9 x 12 square, black or white japanned, convex.
5 8 x 10 square, black or white japanned, convex.

1 10 x 14 soapstone border. Soapstones.
1 9 x 12 soapstone border.
1 8 x 10 soapstone border.................................. **40.00**

3000 hard-burned brick @ $8.50 per M	$25.50	Mason's Materials.
2 barrels common lime @ $1.20	2.40	
2 cartloads sharp sand @ $1.25	2.50	
1 bushel plaster	.60	
		31.00

6 days mason and helper, setting furnace, registers and soapstones	36.00	Labor.
6 days tinsmith and helper, putting up all pipes	30.00	
		66.00

575 hard-burned brick @ $8.50 per M	4.89	Cold-Air Supply.
1 cartload sharp sand	1.25	
¼ barrel common lime	.30	
1 barrel cement	1.40	
59 square feet 2-inch common bluestone	14.75	
Damper and frame of wrought-iron	1.00	
2 days mason and helper building duct and shaft	12.00	
1 day laborer excavating	2.00	
1 man-hole door for shaft	1.75	
		39.34

30 pounds refined galvanized sheet-iron. Hot-Closet.
Screws and washers.
3 wrought-iron shelves.
1 day tinsmith lining closet **9.25**

Brass wire for hanging cellar pipes	.75	Sundries.
Screw-hooks for hanging cellar pipes	.20	
Asbestos for wrapping pipes	1.50	
Furnace cement	.20	
		2.65

Total.................................. **$517.08**

ESTIMATE OF COST OF VENTILATION.

Tin Pipes and Fittings. 24 feet 4 x 12 tin pipe, 3 heads, 1 collar, 1 Y.
3 feet 4 x 10 tin pipe, 2 heads.
4 feet 4 x 9 tin pipe, 2 heads, 2 elbows.
29 feet 8 x 9 tin pipe, 4 elbows, 2 collars.
20 feet 7 x 9 tin pipe, 3 elbows, 3 collars.
8 8 x 10 tin boxes.

Ventilators. 2 9 x 12 square black or white japanned ventilators.
3 8 x 12 square black or white japanned ventilators.
3 8 x 10 square black or white japanned ventilators.
5 8 x 10 square black or white japanned ventilators, convex.

Cords for two ventilators, kitchen and billiard-room $34.33

Labor. 2 days tinsmith and helper, putting in pipes and setting ventilators. 10.00

Total... $44.33

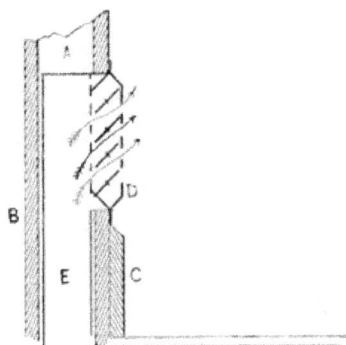

Fig. 8.—*Elevation Showing Register-Opening.*

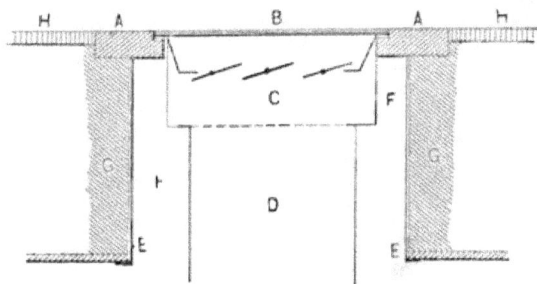

Fig. 6.—*Vertical Section through Wall Register.*

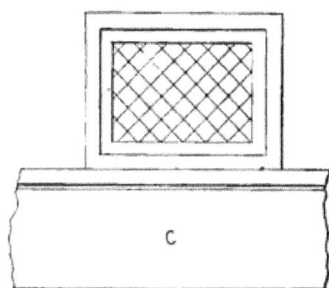

Fig. 9.—*Vertical Section through Floor Register.*

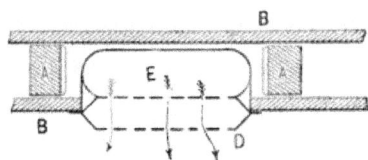

Fig. 7.—*Horizontal Section through Register.*

HOT-AIR SYSTEM.*

BY J. J. WILSON.

GENERAL REMARKS.

Since the days of our first parents man has exercised his ingenuity to obtain protection from the rigors and discomforts of cold. To properly appreciate the progress that has been made in the art of house-heating we must turn to history. Warming apartments by means of hot air was practiced by the early Romans, who erected small furnaces under-ground from which by means of earthen tubes or pipes of different sizes they conveyed the heat to all the rooms of the house. These tubes were invisible, being built in the thickness of the wall, and were connected to that part of the furnace which joined the wall of the house. Through these tubes ascended the heat, which was spread equally and warmed the rooms of the house. Prior to 560 years ago there is no record of a chimney having been used in Europe. They were introduced during the reign of Richard II. Previous to that time our Saxon ancestors, like the red men of America, built fires in the middle of their habitations and provided an opening for the escape of smoke. Then came the open fire-place of monstrous proportions, erected in the center of the house. Economy and comfort have banished these wood-devouring monsters, and it our ancestors could wake after their long nap, as did Rip Van Winkle, they would be compelled to admire the modern open fire with its movable grate, artistic iron linings, art tile and elegant wood mantel with plate mirror.

Heating by hot-water circulation is supposed to have been invented by Bonneman in 1777, and the art of heating by steam was invented by James Watts in 1785, but a practical application of these two systems was not accomplished until some years afterward. The origin of the stove is accredited to Germany, Switzerland and the more northern countries. The modern hot-air furnace is a decidedly American institution. For buildings so constructed that hot air can be conducted by pipes direct to the rooms to be heated the furnace is the best. It is economical in consumption of fuel, its drafts and temperature easly controlled and its ability when in

Margin notes: A Vent Heaters. Chimneys. Open Fire-Place. Hot-Water Heating. Steam Heating. Stoves. Furnaces.

* From *The Metal Worker*, November 2, 1889. Copyrighted, 1889, by David Williams.

operation to furnish a bountiful supply of pure warm air cannot be overestimated. Of late years a number of furnaces, of good construction, have been introduced, and competition has so lowered the price that they can be purchased by the average householder.

Selecting a Furnace.

In selecting a furnace, one should be procured capable of heating the building thoroughly in the coldest weather, without being compelled to overheat its conducting surfaces. The location and exposure of buildings are so widely different that it is best to deal only with those who are known to be practical in the art of heating. This will not only secure a furnace of proper size to heat the given space but the proper locating, setting and arrangement of hot-air pipes and cold-air duct, which is of the utmost importance. If these details are not attended to in a practical manner, the necessary amount of hot air cannot be obtained without a considerable

Parts of the Furnace.

waste of fuel. All furnaces should have a large grate surface with the bars of same not too close together, which will allow the free admission of air, so necessary to perfect combustion; the fire-pot should be much larger in diameter than the grate and its depth not excessive ; a broad and shallow fire-pot presents a large fire-surface for the heat to radiate from. There should be a dust-flue, so that when the grate is shaken the dust from the ashes will not spread over the cellar and its contents. Provision should be made for easy cleaning of the furnace and its radiators, if it has any, without being compelled to dismount or disconnect the hot-air pipes or their adjuncts, except it be the smoke-pipe.

The Vapor Pan.

As for the vapor-pan, I cannot consider it but in the light of a hobby, for it has been demonstrated definitely that pure air is not deprived of its moisture when passed over a moderately heated surface, such as it would encounter in a hot-air furnace. If the moisture were extracted where would it go? If one will make a calculation of the amount of water that is evaporated from the water-pan, and then compare this with the amount of air which passes through the furnace in the course of a day, he can then see how absurd is the claim made for the sanitary effects of the vapor-pan. By the use of a hygrometer the difference of moisture contained in the air before and after passing through the furnace can be easily ascertained, and it will be found there is no difference. I hear some of my brothers in the trade say the people demand them and must have them. In such a case I recommend that where it is possible the water-pan be furnished automatically with water, from the house supply, by means of a float and ball-cock. This will prevent the use of the vapor-pan by servants as a receptacle for cold victuals,

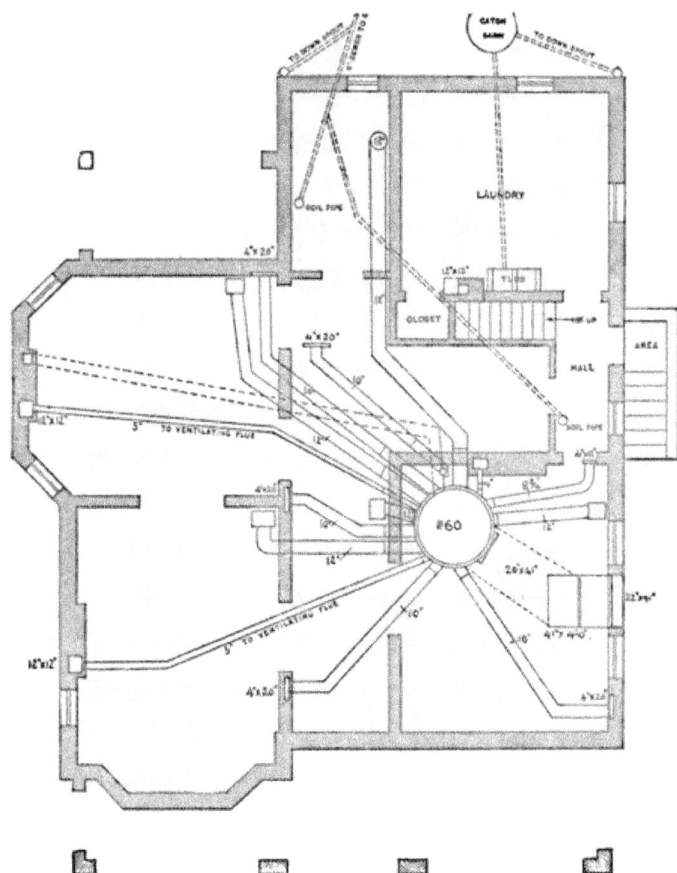

Cellar Plan.—Scale, 1-12 Inch to the Foot.

FLOOR PLANS ACCOMPANYING ESSAY OF J. J. WILSON.

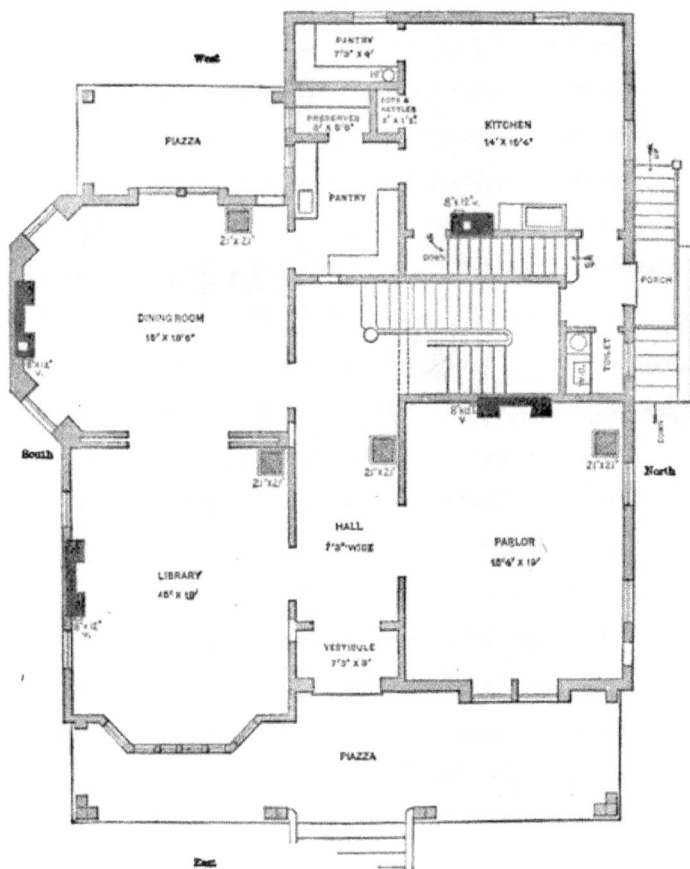

First-Floor Plan.— Scale, 1-12 Inch to the Foot.

FLOOR PLANS ACCOMPANYING ESSAY OF J. J. WILSON.

candles and rat-poison. As to other patent features in different furnaces, the use of them is a matter of taste on the part of dealers—fostered by their experience in erecting and selling them.

HEATING.

It is to be assumed that I have secured the contract for heating the dwelling in the House-Heating Competition of *The Metal Worker*. I propose to call the architect's attention to the space and provisions to be made for the erecting of the hot-air pipes in the studding and walls so that he can notify the contractor in time. It will be noticed in general dimensions of plans that the joists are 16 inches between centers, and as the joists are usually 3 x 10 inches this would leave 13 inches in the clear between the joists. The studding, 2 x 4 inches, is 16 inches between centers ; this leaves 14 inches in the clear between the studding. As my wall-pipes up at the second story are 4 x 20 inches, I must have space between the joists at least 21 x 5 inches. The joist should set 1 inch away from the pipes. This is easily accomplished by putting in a trimmer, which is a short joist, mortised between joist at the width needed. The joist which is cut to make room for the pipe mortised into the trimmer.

The same general plan is pursued for the floor registers. The framing timber or outside joist which lay upon the walls is generally 4 x 6 inches wide. The outside studding of frame is fastened to this timber, either nailed or mortised; the joint which binds upon this timber is generally cut so that it rests upon this frame and also rests upon the wall. The joist being 10 inches high and the framing timber 4 inches high when the joist is cut out there will remain 6 inches resting upon the frame. If I were using a square 4 x 20 inch elbow this would do, but as space must be had for a 10-inch pipe to enter the elbow or starter I will cut away this frame in a slanting direction and make my starter to suit as seen in Fig. 1. I propose to have ready before dwelling walls are built the necessary wall collars for the hot-air pipes in cellar to pass through. They should be made the thickness of the wall, and when angled space allowed to bevel collars. Their diameter should be ½ inch larger than the pipe which passes through them. This makes a neat job, does not weaken the wall and admits of the pipes being taken down without disfiguring the wall. These should be built in the cellar walls as they are being erected by the mason. I propose to get ready at once all the material needed for the wall-pipes, so they can be put in position as the studding is being erected. I propose to have

Space for Hot Air Pipes

Setting Registers.

Reducing Elbow.

Wall-Collars.

Wall-Pipes

the flues in chimneys changed to suit system of ventilation I intend to use.

As soon as the roof is on the building and studding is up I propose to place wall-pipes in position and lath them in. Then to excavate space required for the furnace, cold-air pit, cold-air box and foundation of cold-air storage-room. *Brick Work.* The brick-work built up and covered in as the cellar floor is to be cemented and all the work can be cemented in at one time. The frame for the cold-air opening should be built in cellar wall when they are being put up. The furnace and the cold-air storage-house which I intend to use can be put up after dwelling is plastered.

Furnace. I propose to use a No. 60 portable furnace with cast ash-pit, cast fire-pot, cast magazine or feed section and enlarged steel radiator. The dimension of the grate-surface is 30 inches; diameter of fire-pot, 34 inches; depth of fire-pot, 14 inches; diameter of steel radiator, 57 inches; hight, 14 inches; hight of castings, 68 inches; size of smoke-collar, 8 inches; diameter of lower casings, 60 inches; diameter of upper casing, 68 inches. The grate shakes and dumps, with space to clean out clinkers. My preference and reasons for using this size of furnace are that there are a large number of pipes to be run from it, requiring a furnace of large capacity to thoroughly fill them all. Besides, it admits of all the pipes being taken from its head without being compelled to run them from space in a line with the ash-pit front and feed-door, as there is very little current of air over them, the pit and door being an obstruction. It is fully ample to heat the space required in the coldest weather.

Location of Furnace. I propose to locate the furnace in the cellar under the parlor in the southwest corner of this division. This places the furnace close to the north side, which is the coldest portion of the dwelling and the part that should be most favored with short and direct pipes. It also admits of a short cold-air box without many bends, it being taken from the north side, which is the best point to bring cold air from, notwithstanding controversy among some furnace men to the contrary. It is true that strong winds come from this direction, and without special provision is made will cause the air to come out of the registers in puffs and only partly heated. This I propose to overcome in the construction of my cold-air storage-room, which I will explain further on. The point at which I have located the furnace is convenient for unloading and storing fuel, has plenty of light and is convenient to division wall, allowing the majority of the dampers to be operated in the cellar passageway without being compelled to go in division in which the furnace is set.

I propose to use the best quality IX charcoal bright tin- Hot-Air Pipe plate for all the work requiring tin. I propose to run from the cellar to the second floor for all the rooms 8 x 20 inch hot-air pipe set in the studding with the exception of the sewing room and chamber adjoining, as they are connected from first to second floor with a 10-inch pipe run through the closet adjoining the kitchen. From the second to the third floor pipe for all the rooms will be 4 x 12 inches, connected to the 4 x 20 inch pipes in the second floor and their flow of air controlled by the flue-damper located just above the register in floor below.

To make the 4 x 20 inch pipe requires two sheets of 20 x Construction of Pipes. 28 inch tin—one sheet cut 27¾ x 20 inches and one sheet cut 21 x 20 inches; this allows for edges. The waste piece left, 6¾ inches, will make the gore for the 4 x 12 inch tin pipe. Notch the tin at both ends for the grooving-edge and lay off proper space for squaring pipe to size needed. At each corner cut in ⅜ inch deep for double-seaming the pipe together, as all pipe put in the studding should be locked together in as long pieces as possible. Edge the tin in the brake for grooving. At one end of pipe turn up a grooving-edge, then unlock the folder or brake, reverse the tin and turn up another edge on the opposite end of the pipe, turn up an edge at right angles to the sheet, bend tin at the spaces laid off in square pipe-former or use the flat end of beakhorn stake; groove together. To connect joints pin the edges together and double-seam. Fig. 2 explains position of pipe before double-seaming and Fig. 3 the position in the brake. Two braces are required at each end to be soldered in 4 x 20 inch pipe. They should be placed 7 inches apart, nearly dividing the pipe into three equal spaces, and 5 inches from each end. This stiffens it and prevents it from bulging in.

The starters, which connect cellar-pipes with wall-pipes, Starters. are made as shown in Fig. 1. Headers on pipe with box for Headers. register are made the same as the pipe, except that they are connected with a long slip-joint, well soldered. Lay off size of opening for register on the header. Draw a diagonal line from the upper right-hand corner to lower left-hand corner of opening; draw same line from opposite side; cut through these lines. Bend out at right angles to pipe ⅞ inch, which is the thickness of the plaster; cut off what projects beyond this; solder in the corner-pieces. Double-seam on the head and fasten in 4 x 12 inch collar for the 4 x 12 inch tin pipe if the heat is to be carried to the room above.

To make the 4 x 12 inch pipe cut one sheet tin 27¾ x 20 4 x 12 Pipe. inches, and gore 5 x 20 inches; this is cut out of waste left by 4 x 20 inch pipe; pursue the same course in making pipe and boxes as before described. In the 4 x 12 inch collar, close to top of register in 4 x 20 inch pipe, place a 4 x 12 inch flue-damper, with ₇⁄₈ inch rod, and when lathing punch a hole in a lath large enough for this handle to run through, fasten in the rod and bring it through lath, in a line, so it will work easily ; this will prevent rod from being bent out of line.

Lining Studding. Before putting pipe in the studding line each side of studding with tin cut 4 inches wide and nailed against the thickness of same. In the two runs of hot-air pipe in the outside wall against sheathing line the three sides of opening in the same manner, using tin cut 5 x 28 inches for the sides, with a 1-inch edge, bent up at right angles to the tin ; this rests against the sheathing. Nail 20 x 28 tin ; **Pipe-Supports.** this will cover space 4 x 21 inches nicely. Tin bands about 3 inches wide should be soldered on the face of pipe, and when it is placed in position, nailed against the studding; these should be placed about 4 feet apart.

Pipe-Covering. I propose, after studding-pipe is in position and before lathing, to wrap the pipe with asbestos sheathing, the weight of which is about 6 pounds to 100 square feet and width 42 inches. Cut off the proper length to admit of a good lap, letting the 42-inch width run perpendicular with the pipe ; tie it neatly around the pipe with a good quality of annealed tin-coated wire. The covering for register-boxes is made by cutting the sheathing in the same way as described for laying off register-box in headers. Slip it over the opening and tie above and below box. In the outside studding against sheathing the pipe can be wrapped with asbestos before being placed in position.

Lathing. For lathing use No. 27 black sheet-iron, cut 2½ inches wide and 23 inches long. Turn up two ¼-inch edges, both on the same side, mash them, and then turn up another ¼-inch edge ; this stiffens the lath and prevents it bending ; punch a hole in each end for lathing-nail. In putting up the laths they should touch each other, so that a perfect key is made for the plaster to fasten to (see Fig. 4). Nail up on both sides of studding, and around the register-boxes form a square that will slip over box and extend to studding on both sides. This is accomplished by taking two laths of full length and riveting between them pieces a little longer than the hight of the register-box (see Fig. 5). This is good construction and forms a perfect lath for plastering. Where the pipe runs near joist and base-board roofing slate should be nailed over the studding instead of the laths. Near the joist it is easier to get at and at the base-board makes a better finish.

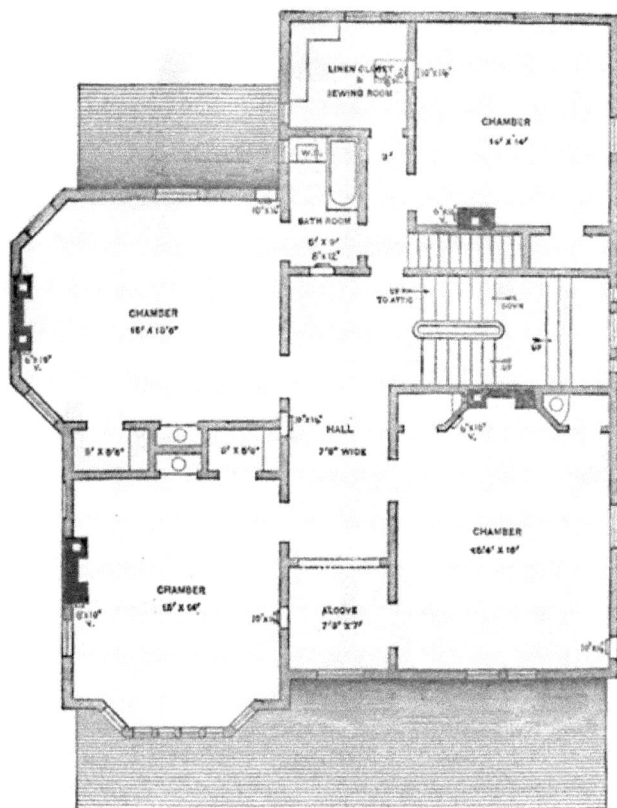

Second-Floor Plan.—Scale, 1-12 Inch to the Foot.

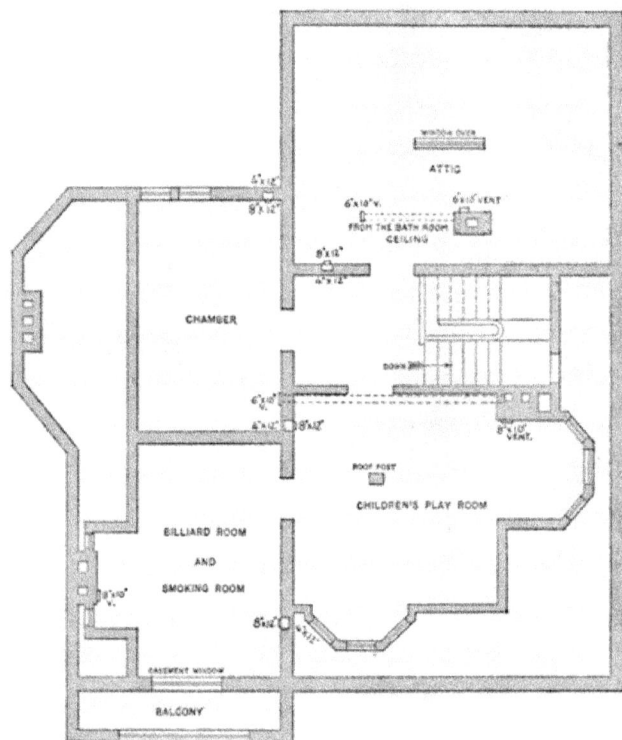

Attic Plan.—Scale, 1-12 Inch to the Foot.

I propose to run a 10-inch pipe through the closet adjoining the kitchen, to connect with a boot or pipe to sewing-room and chamber over kitchen; the lower end of 10-inch pipe is extended at cellar to 12-inch pipe. The pipe running between the joists will be 8 x 20 inches and taper to 4 x 20 inches, connecting by means of an elbow with a small circle to the 4 x 20 inch pipe in the studding (see Fig. 6). Two register-boxes are brought out at opposite sides, one 8 x 10 inches for the sewing-room, the other 10 x 14 inches for the chamber. There is a division-piece put between them which should extend 4 inches below the bottom of the 10 x 14 inch register-box (see Fig. 7). The pipe running between the joists is to be covered with the asbestos sheathing neatly tied. Tin is to be nailed between the joists over the pipe ½ inch below face of of the joist. The sides and bottoms of joists to be covered with tin. Under the pipe on the ceiling of closet nail iron laths. This makes a strictly fire-proof construction. *(margin: Pipe to Double Register.)*

I propose to have a window-frame set in the cellar wall, size 22 x 41 inches, on the north side of the dwelling, as indicated on the plans. To build against this a storage-room for cold air, said room to be built from cellar floor to joist and divided in the middle by a partition extending from the roof to 18 inches below window-frame ; this is to be built across the width of the house to form a break for the wind. The dimensions of this storage-house are 41 inches wide, 48 inches deep and 8 feet 6 inches high, to be built of tongue-and-groove flooring with the studding on the outside, presenting a smooth surface for flow of air ; this storage-room is to be built upon a substantial foundation of brick and cement with a slate floor, the depth of foundation to conform to space of cold-air box, 41 inches wide and 20 inches deep. In the front division of the room strips and a door are to be built to admit of a frame covered with ladies' veiling being taken out and put in at pleasure, in order to clean or renew. This veiling is positively known to be the best fabric that can be used for catching dust and preventing it entering by circulation the rooms connected with the furnace. The frame should be placed 2 feet above the floor-line of cellar. In building the storage-room the tongue-and-groove lumber should be painted with thick white lead as it is put together, in order to make the room as nearly air-tight as possible. This will prevent the air from being deviated from the cold-air box and entering the cellar. *(margin: Cold-Air Storage-Room.)*

There is to be a damper 23 x 41 inches hung on hinges at its top and fastened to sill of window-opening (see Fig. 9), a staple being fastened near the lower end. To this fasten a small chain, run it up to ceiling and over a pulley-wheel in *(margin: Cold Air Damper.)*

the top of room; run the chain through a hole in the partition and through the side of room to the cellar, from which it is to be conducted up to the hall and the end of chain hung upon a hook. When the wind is blowing very strong from the north or west the box can be partially or entirely closed and air taken from the cellar. There is to be a door 3 x 2 feet (see Fig. 8) cut in front side of room next the furnace, the door to be hung on hinges at its base and fastened at the top by means of a wooden button. When the damper at the window is closed this door can be opened and air taken from the cellar. I do not approve of air being taken from the hall of a house, as foul or impure air is drawn from the floor-line, reheated, and delivered to the different apartments, to be breathed over again. The opening of window-frame for cold air is to be covered with galvanized wire-cloth, ¼-inch mesh, to prevent the entrance of leaves, dirt, &c. The slate upon which the foundation of the storage-room is built is 1 inch thick; it is damp-proof, and allows the air to pass over it without friction. Fig. 8 shows construction of storage-room and Fig. 9 construction of the damper at window-sill.

Cold-Air Pit.   The cold-air pit under furnace is to be built of brick and cement. Its dimensions in the clear are 20 inches deep and 60 inches in diameter, smoothly plastered inside with cement, and a brick bottom laid in the same material. A brick cold-

Cold-Air Box. air box is to be built from the storage-room to cold-air pit, connecting with both, the bottom to be of slate 1 inch thick and 49 inches wide. On this slate the brick are to be laid in cement 20 inches high and 41 inches wide, which are dimensions of the box in the clear. The top of cold-air box is to be covered with slate slabs 1 inch thick and laid so that they will be 1 inch below the floor-line; this will admit of their being cemented over. All of this work can be covered at the same time cellar floor is cemented. The walls of all the above work are 4 inches thick and the work laid with the best Portland cement mixed as follows: 1 part cement to 2 parts sharp sand. This I consider to be first-class construction in every respect.

The Furnace.   The furnace is built with a double casing of No. 24 galvanized iron. The inside casing to be covered with asbestos sheathing on side next the furnace and fastened with 1½ inch galvanized sheet-iron washers, riveted through to casing. The top of furnace to be flat, with a 4-inch sand-rim, the inside of said top to be covered with asbestos sheathing and a convex top to be riveted over this. The top of furnace to be covered inside of sand-rim 1 inch deep with asbestos cement felting, the other 3 inches to be filled in with sharp sand. The joists above the furnace are to be covered with No. 24 galvanized iron, forming a square of about 8 feet. The work finished in this manner is strictly fire-proof and it is impossible for even a small amount of heat to be wasted in the cellar by radiating through casings. Construction of this kind saves a considerable amount of fuel and allows the pipes leading from the furnace to get the full benefit of all the heat radiated from the fire. The cast sections of furnace are mounted with kaolin ; the steel radiator and doors with asbestos furnace-cement. The base of the furnace rests on a level with the cellar floor, which I prefer to sinking it, for when the furnace is set in a pit it is hard to keep the pit clean and it is inconvenient to take out the ashes from under the grate unless a very large space is allowed.

The parlor, library, dining-room and hall on the first floor are connected to the furnace by 12-inch pipes. All wall-pipes above the first floor, with one exception, are connected to furnace by 10-inch pipes. The sewing-room and chamber adjoining it are connected from the furnace to the closet next the kitchen by a 12-inch pipe. The water-closet on the first floor on the north side of building is connected to the furnace by 8¾-inch pipe. These sizes, with proper elevations, are ample to heat the space designated on the plans and to the temperature specified. The Cellar-Pipes.

The parlor register is located in the northwest corner of parlor. It is a floor register, 12 x 15 inches, with iron floor-border ; this is covered with a 21 x 21 inch pedestal register-stand with marble top and finished in gold bronze. This permits a neat finish to be made by the carpet around the pedestal, protects the inside register at the floor-line from outside currents of cold air and is an ornament to the parlor. The library register is located in the northwest corner of the room and is of the same size and finish as the parlor register. The dining-room register is located in the north-west corner of room, space being left for the door leading to pantry to open. My reason for not placing it on the north side of room against the wall is that this space is usually reserved for the side board. The size and finish of register is the same as the parlor register. The hall register is located against the north side of hall in order to protect it from cold currents coming from that direction. The size and finish is the same as the parlor register. The register in the water-closet on the first floor is located in the east end of room, it being the only space available for it. The size is 8 x 10 inch square register, set in the wall. First-Floor Registers.

On the second floor in the chamber over parlor the register is placed in the wall at the north side near the east front, Second-Floor Registers.

protecting it from all currents of air from the north ; its size
is 10 x 14 inches square register. The register in the chamber over the library is located in the north wall near the east
front and is a good location, for the same reasons given before.
The size and finish are the same as in the room over the parlor.
The hall register on the second floor is located on the east
side of hall and is placed there because this wall is in a line
with the studding in the wall of the children's play-room,
which admits of the pipe being continued up to said room.
The size of register is 10 x 14 inches square. The register in
chamber over the dining-room is located at the northwest
corner of the room and is a good location, for reasons given
before. The size of register is 10 x 14 inches square. The
register in the bath-room is located in the east wall of said
room. This is the only place one can be put without interfering with the fixtures of the room. It is placed so that if
the door leading to the hall on the one side and the door leading to the chamber on the other side are opened they will not
strike against the face of the register, which would be apt to
blister and discolor them by its heat. The size of the register
is 8 x 12 inches square. The sewing-room and the chamber
over the kitchen are both connected to the one pipe. In the
chamber the register is located in the south wall of room. This
is about the best location that can be secured under existing
circumstances ; the northwest corner would have been the
best, but the distance in a line from the furnace is too grea
and would also take up considerable room in the laundry. In
the location selected the sewing-room can be heated from the
same pipe ; it being of a large capacity and not running
any higher up it will furnish a sufficiency of heat for both
rooms. The size of the register in the chamber is 10 x 14
inches. The size of the register in the sewing-room is 8 x 10
inches.

The register in the children's play-room is located in the
south wall of the room. This room is not exposed much to
the wind and therefore this register will furnish all the heat
needed for the required temperature. The size of the register
is 8 x 12 inches square. The register in the billiard-room is
located in the north wall near the east front, which is a good
location ; besides, being located near a corner of the
room, it is not liable to blister or discolor the billiard-table.
The size of register is 8 x 12 inches square. The register in
the chamber adjoining the billiard-room is located in the
northwest corner of the room and is a good location. The
size of register is 8 x 12 inches square. The register in the
attic is located in the east wall, being placed there because
this wall is in line with the studding of bath-room, permitting

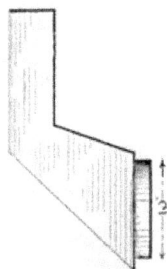

Fig. 1.—Reducing Elbow.

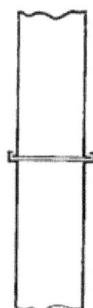

Fig. 2.—Before Double-Seaming.

Fig. 3.—Position in Brake.

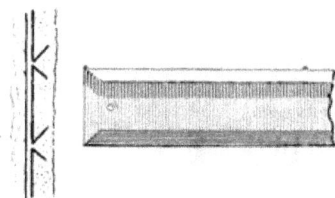

Fig. 4.—Sheet-Iron Lath and Method of Fastening.

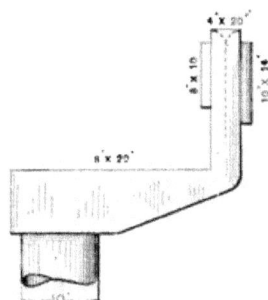

Fig. 5.—Laths Around Register-Box.

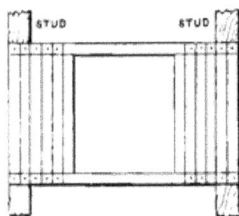

Fig. 6.—Piping to Two Registers.

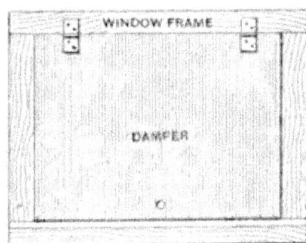

Fig. 9.—Damper at Window-Sill.

DETAILS OF CONSTRUCTION ACCOMPANYING ESSAY OF J. J. WILSON.

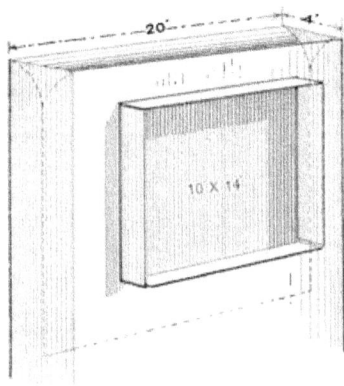

Fig. 7.—Register-Box.

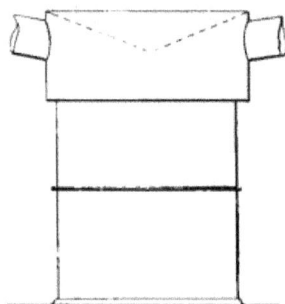

Fig. 10.—Elevation of Furnace.

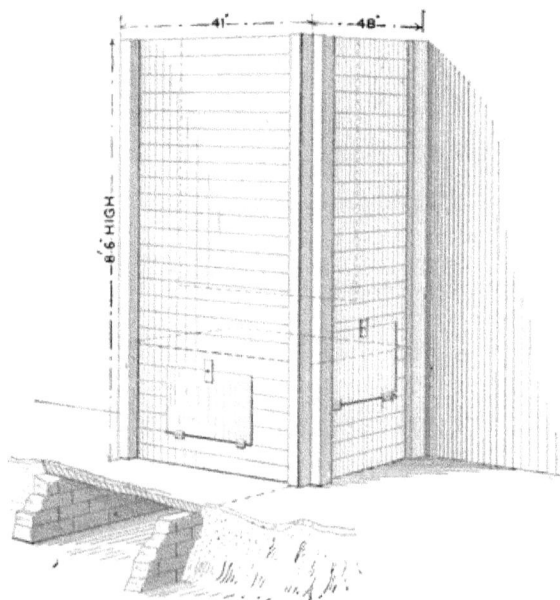

Fig. 8.—Vertical Part of Air-Box.

DETAILS OF CONSTRUCTION ACCOMPANYING ESSAY OF J. J. WILSON.

the pipe to be extended from the bath-room to the attic.
Although this room is on the northwest corner of the dwelling it is pretty well closed in, there being no windows in the
room except at the ceiling-line; for this reason it is easily
heated.   The size of the register is 8 x 12 inches square.

The 12-inch pipe used in the cellar is made from two *Pipe Con-*
sheets of 14 x 20 inch tin, cut 14 x 19⅜ inch, edged and grooved *struction.*
together. Swedge with O. G. swedge 2 inches from the end,
crimp pipe for small end and solder together. The 10-inch
pipe is cut from one sheet 27⅜ inches and a gore 4½ x 20 inches
wide ; follow the same plan as for 12-inch pipe. The 8¾-inch
pipe is made from one sheet, cut 27¼ x 20 inches ; follow the
same plan as for the 12-inch pipe. The 12 x 15 register- *Register-*
boxes are cut two pieces 16¼ x 27 inches and two pieces *Boxes.*
12⅞ x 27 inches ; on the long pieces turn up two edges, both
on the same side of tin ; on the same pieces turn up two
edges, both on the same side of tin and at right angles. The
four corners of the tin should be notched to allow an edge for
the bottom and ¼-inch edge for the top of box to rest on
the register-border ; turn up edge on bottom and top of box.
Pin a large and small edge together and double-seam them ;
pursue the same plan until box body is finished ; cut out the
bottom 13½ x 17 ; turn up a ⅜-edge all around ; pin it into
the register-box body and double seam.

The different-sized oblique elbows are made from one joint *Elbows.*
of tin, cut at proper angle before grooving. The four-piece
right-angle elbows are cut the same size in circumference as the pipe and cut at proper angle before
grooving. The patterns for these angles are obtained in the
following way : Draw the elevation of the correct hight, diameter, length and pitch of elbow. Divide the circle of elbow
into four parts ; now divide the elevation into any number of
spaces ; number them 1, 2, 3, &c. Construct a parallelogram,
equal in length to the circumference of the elbow, draw a line
through the center, dividing equally the length of circumference. Lay off on each the same number of spaces there are
in the elevation, numbering them 1, 2, 3, &c.; lay a square
upon the lines in elevation, and at right angles draw a line
from No. 1 in the elevation to No. 1 in the parallelogram.
Pursue this course with the lines at their angles in the elevation, carrying them across to the corresponding line in the
parallelogram ; a line drawn through the point where lines
intersect on the parallelogram will be the pattern.

There will be three 12-inch goose-neck elbows used in this
job. The reason for making the floor register-boxes 27 inches
deep is that this will permit pipe being put into them with
out using extra elbows, as pipe-holes are cut in sides. There

will also be used in this job three 12-inch oblique elbows and seven 10-inch oblique elbows.

Regulators. I propose to operate draft of furnace, check-damper in smoke pipe and damper in cold-air box by means of small chains attached to these fixtures and carried over wheel pulleys up into the hall so they can be operated without coming into the cellar.

All the hot-air pipes lead from the side of the furnace, Fig. 10 being an elevation of furnace. The top of the pipes *Elevation of* at the furnace will be 81 inches from the floor-line of cellar. *Cellar Pipes.* The hight of the cellar in the clear is 102 inches. Taking off 4 inches, the space between ceiling and top of pipe where it enters the wall-pipe and register-boxes, there will remain 98 inches, leaving 17 inches for elevation. My two longest pipes—a 5-inch ventilating hot-air pipe running to ventilating-flue of dining-room and a 12-inch pipe running to closet adjoining kitchen—have an elevation of 1 inch in each 18 inches of length, the longest of the other pipes about 1 inch in each 14 inches of length, and the balance 1 inch in each 11 inches in length and from that up.

Pipe Area. • The area of the pipes leading from the furnace are as follows

			Inches.
5 12-inch pipes,	113.097 each		565.485
5 10-inch pipes,	78.540 each		392.700
1 8¼-inch pipe,	60.132..		60.132
2 5-inch pipes,	19.635 each		39.270
1 4-inch pipe,	12.566		12.566
			1070.153

Three-quarters of this space for cold air.................... 803.615
Size of cold-air box 20 x 41 inches.......... 820

These measures are obtained by multiplying the square of the diameter by 0.7854, as follows : 12-inch 12 × 12 = 144 0.7854 = 113.097. This rule applies to all sizes of pipe.

Pipe-Sheathing. All the pipes in the cellar, including the smoke-pipe, are to be wrapped with asbestos sheathing ; the sheathing should be cut of such a length as to allow a good lap around pipe; place it around the pipe with the 42-inch width running horizontal with the pipe; tie neatly around the pipe with good annealed tin-coated wire. The cellar pipes should be well soldered *Pipe Supports.* together and their weight supported by strong wire neatly wrapped around the pipe and fastened to the ceiling or joists. Wherever any of the pipes come close to wood-work in the cellar it should be lined with tin.

HEATING ESTIMATE.

I will now consider the cost of material and labor in constructing and setting up the furnace, hot-air pipes, registers, cold-air duct and storage-room and cold-air pit under the furnace :

No. 60 furnace, with mounted steel radiator.......................... .... $128.25

Bottom section of outside casing for furnace, No. 24 galvanized iron, diameter 60 inches, hight 20 inches; cut one piece 20 x 94 inches and one piece 20 x 95 inches; two grooves allowed for; weight about 26½ pounds.

Middle section, diameter 60 inches, hight 24 inches; cut one piece 24 x 94 inches and one piece 24 x 95 inches; two grooves allowed for; weight about 31⅜ pounds.

Radiator section 20 inches by 17 feet 11¼ inches; cut two pieces 7 feet 11 inches each and gore 2 feet 1⅞ inches; hight 20 inches; three grooves are allowed for; weight 30 pounds.

Top or pipe section 14 inches, length 215½ inches; cut same as for radiator section; weight 21 pounds.

Size of flat top cut 69 inches; weight about 17¼ pounds.

Convex top cut 69 inches; weight about 17¼ pounds.

Inside casing, No. 24 galvanized iron, bottom section, diameter 57 inches, hight 20 inches, length 15 feet ½ inch; cut one piece 7 feet 11 inches and one piece 8 feet 1½ inches, the two grooves are allowed for; weight about 25¼ pounds.

Middle section, diameter 57 inches, hight 24 inches, length 15 feet ½ inch; cut same as for bottom section; weight about 30¼ pounds.

Radiator section, diameter 65 inches, hight 20 inches, length 19 feet ⅜ inch; cut two pieces 7 feet 11 inch each and gore 3 feet 2⅜ inches; weight about 30½ pounds.

Total, 229 pounds of iron, @ 6¢............................... 13.79

For sand-rim 4 inches wide, cut two pieces 7 feet 11 inches each and gore 5 feet 2 inches; weight about 7 pounds, @ 6¢......................... .42

Wire for sand-rim, about... .10

Labor for casing furnace, 1½ days for hand @ $3 per day and helper @ $2 per day.. .... 7.50

Total for furnace $150.06

Five lengths of 4 x 20 inch tin pipe, each length about 14 feet 4 inches and will take one sheet tin cut 27¾ inches and one sheet cut 21 inches; 18 sheets of tin, 20 x 28 inches, for each length, @ 16¢ per sheet.......... 14.40

Five starters, 4 x 20 inches, with splayed bottom and with 10-inch hole; 16 sheets 20 x 28-inch tin, @ 16¢.. .... 2.56

One 14 x 20 inch circle elbow; 3 sheets 20 x 28-inch tin, @ 16¢............ .48

One 4 x 20 inch head, with 10 x 14-inch register-hole on one side and one 6 x 10 register-hole on the opposite side, with one division-piece between them; 3 sheets tin, @ 16¢ each.................................... .48

Four lengths of 4 x 12 inch tin pipe, each length about 10 feet 3 inches, cut from one sheet 27¾ inches, and the gore, 5-inch wide, is cut from the waste of the 4 x 20 inch tin, seven sheets in each length; 28 sheets of tin, @ 16¢ each... .... 4.48

One starter, 4 x 12 inch, with splayed bottom, with 8½-inch hole; three sheets 20 x 28 inches, @ 16¢ each.............. .... .48

Three sheets of tin for reducing collars, 4 x 12 inches, and top of 4 x 20
    inch headers, 20 x 28 inch tin, @ 16¢ each........................ **$0.48**
    All the 4 x 20 inch and the 4 x 12 inch headers are calculated for in
    the regular run of the pipes.
Labor for above work, three and one-half days, @ $3 per day............. 10.50
Solder, 2 pounds, @ 20¢ per pound................................... .40
67 feet of 12-inch hot-air pipe, cut from two sheets of tin 14 x 19$\frac{3}{16}$ inches,
    each joint when put together measuring 1 foot; 134 sheets of tin 14 x 20
    inches, at 8¢ per sheet... 10.72
90 feet 6 inches of 10-inch hot-air pipe; cut one sheet 27$\frac{1}{8}$ inches and a
    4$\frac{1}{2}$-inch gore, 6 out of a sheet; 71 sheets of tin 20 x 28 inches, @ 16¢
    each.. 11.36
Three 12-inch oblique elbows, one elbow out of two sheets of 14 x 20
    inch, @ 8¢ each.. .48
1$\frac{1}{2}$ pounds of solder, @ 20¢ per pound .... .30
Seven 10-inch oblique elbows cut same as for 10-inch pipe; nine sheets re-
    quired, 20 x 28 inches, @ 16¢ each................................ 1.44
Three 12-inch goose-neck elbows in 3 pieces cut same as for 12-inch pipe;
    six sheets required, 20 x 14 inches, @ 8¢ each..................... .48
One boot or pipe to run between the joists above the closet next the kitchen
    and connecting with 4 x 20 inch pipe in the studding of sewing-room and
    chamber adjoining; boot to have 10-inch collar on bottom; size 20 inches
    long, 8 x 20 inches at one end and 4 x 20 inches at the other end; three
    sheets of 20 x 28 inch tin, @ 16¢ each............................ .48
2 feet 6 inches 4 x 12 inch tin pipe for water-closet on the first floor, with
    8 x 10 inch register-box cut in pipe and a head double-seamed on pipe;
    the gore needed for this pipe is taken from the waste of the 4 x 20 inch
    pipe; two sheets of tin, @ 16¢ per sheet.......................... .32
9 feet 8$\frac{1}{2}$ inches hot-air pipe cut 27$\frac{3}{4}$ x 20 inches; six sheets of tin, 20 x 28
    inches, @ 16¢ each... .96
Labor for above work, four days, @ $3 per day..................... 12.00
Galvanized collars for cellar walls for pipe to pass through; No. 24 iron; six
    12$\frac{1}{2}$-inch collars, 15 inches long, cut 40 inches; about 7$\frac{1}{2}$ pounds, @ 6¢
    per pound... .81
Four 10$\frac{1}{2}$-inch collars, 15 inches deep, cut 33$\frac{5}{8}$ inches; No. 24 iron; about
    7$\frac{1}{2}$ pounds, @ 6¢ ....:....................... .45
Labor for above work, two hours, @ $3 per day..................... .60
Four 12 x 15 register-boxes 27 inches deep; five sheets of 20 x 28 inch tin, @
    16¢ each... 3.20
Labor for above work, three-quarters of a day, @ $3 per day........... 2.25
Five 12-inch galvanized-iron pipe-collars for furnace, No. 24 iron, cut 38 x 8
    inches; five 10-inch galvanized-iron pipe-collar for furnace, No. 24 iron,
    cut 31$\frac{1}{2}$ x 8 inches; one 8$\frac{1}{2}$-inch galvanized-iron pipe-collar for furnace,
    No. 24 iron, cut 27$\frac{3}{4}$ x 8 inches; about 5$\frac{1}{2}$ pounds, @ 6¢............. .33
Labor for above work, three hours, @ $3 per day.................... .90
11 patent spring-dampers for hot-air pipe, @ 25¢ each.............. 2.75
16 joints of galvanized smoke-pipe, No. 24 iron, 8 inches in diameter; about
    4$\frac{1}{2}$ pound per joint, 72 pounds, @ 6¢ per pound... .... 4.32
Three 8-inch smoke-pipe elbows, No. 24 galvanized iron; about 6$\frac{3}{4}$ pounds, @
    6¢ per pound.. .41
Two galvanized wall-collars for smoke-pipe to run through in cellar wall,
    size 8$\frac{1}{2}$ inches in diameter and 15 inches long; about 2 pounds, @ 6¢ per
    pound... .12
Labor, one-half day, @ $3 per day................. .... 1.50

Tin lining for studding, three runs, about 24 feet each run. tin cut 4 x 28
inches; five pieces out of a 20 x 28 inch sheet tin; four and one-half
sheets for two sides of lathing, @ 16¢ per sheet...................... $0.72
One run in outside studding lined on two sides with tin, 5 x 27, with
1-inch edge turned off at right angles to lay against outside sheathing,
and one sheet of tin 20 x 28 tin on the back of opening against the
outside sheathing, about 24 feet; 16½ sheets 20 x 28 inch tin, @ 16¢ per
sheet..   2.56
One run in outside studding lined with 20 x 28 inch tin; length of run about
14 feet 4 inches, lined on two sides with tin cut 5 x 28 and on the back
against sheathing a sheet of tin 20 x 28 inches; about nine and one-half
sheets of tin, @ 16¢ per sheet..   1.52
For lathing three runs of pipe from the cellar to third floor on two sides;
number of iron laths 1140; for lathing one run of pipe in outside wall from
cellar to third story, number of laths 190; for lathing one run of pipe
from the cellar to second story in outside wall, number of laths 108; for
lathing 1 run of pipe to water closet on the first floor, number of laths
28; for lathing horizontal pipe running between joists and pipe running between studding in sewing-room and chamber adjoining, number
of laths 48; total number of laths, 1514; they cut 38 out of sheet; laths
are cut 2¼ inches wide by 23 inches long, and after being edged they are
1½ inches wide by 23 inches long; No. 27 black-iron, 12 pounds to
the sheet, 40 sheets—480 pounds, @ 3½¢ per pound.................. 16.80
Labor for making laths, one and one-half days' work, @ $3 per day........   4.50
Brick-work for cold-air pit under the furnace, cold-air box and foundation
walls for cold-air storage-house, walls 4 inches thick for cold-air pit, 20
inches deep and 5 feet in diameter, 176 brick; 4-inch wall for cold-air
box, 7 feet long and 20 inches high, 160 brick; brick bottom for cold-air
pit under the furnace, 5 feet in diameter, 140 brick; foundation wall on
the west side of storage-house, 2 feet by 20 inches high, 24 brick; east
wall, 4 feet by 20 inches high, 72 brick; extra brick for waste, &c., 25;
number of brick 597, @ $10 per 1000.... ....   5.97
Mason's and brick-layer's time building brick-work, one day and one-half,
@ $6 (mason, $4; helper, $2)..   9.00
1½ barrels of Portland cement, @ $3.50 per barrel.......................   5.25
3 barrels of sand, @ 25¢ per barrel.....................................   .75
Lumber for cold-air storage-house, size 8 feet 6 inches high, 4 feet deep and
41 inches wide, with waste, about 135 square feet, @ 6¢ per foot........   8.10
5 pounds 10d nails, @ 4¢ per pound.....................................   .20
42 feet of 2 x 4 inch studding, @ 4¢ per foot...........................   1.68
Extra lumber for door and frame, about 25 feet of flooring, @ 6¢ per foot.. 1.50
Ladies' veiling for frame in storage-room, about 7 yards, @ 25¢ per yard... 1.75
White lead for painting tongue and groove.............................   .25
Carpenter's time building room, two days, @ $3 per day....... ... .   6.00
2 pair hinges, for the doors, @ 15¢....................................   .15
5 pounds of lathing-nails for studding, to fasten iron laths, @ 4¢ per pound.. .20
Four 4 x 12 flue-dampers, @ 40¢ each..................................   1.60
4 short rods, ⅜, with thread, @ 6¢ each.............................   .24
4 protection caps, @ 8¢ each..   .32
4 nickel knobs, @ 5¢..   .20
10 sheets of tin for covering around joists, &c., @ 15¢ each.........   1.60
Wall tube, 4 x 4 inch galvanized iron, for chains from the cold-air damper,
draft-door in ash-pit and smoke-pipe check-damper to run through cellar
wall up into hall, about 1½ pounds, @ 6¢...........................   .09

One package of plaster-paris, 15¢ $0.15
82 square feet of slate, 1-inch thick, for bottom of cold-air storage-house,
   bottom of cold-air box and top of cold-air box, @ 25¢ per square foot.. 20.50
Four 12 x 15 inch registers and borders, @ $2.52 each.................... 10.08
Five 10 x 14 inch registers, @ $1.73 each.......................... .... 8.65
Five 8 x 12 inch registers, @ $1.20 each............. .... 6.00
Two 8 x 10 inch registers, @ $1.05 each.............................. 2.10
Three 21 x 21 inch pedestal-registers, with marble-top, door and shelf at-
   tachment and gold bronzed, @ $19.50 each 58.50
One 21 x 21 inch pedestal-register, color chocolate tipped with gold, and
   door and shelf attachment, @ $18................................ 18.00
Three pounds of tinned wire, for tying asbestos on pipes, @ 15¢............ .45
50 roofing-slate, for covering joists and placing between base-board and
   pipe, @ 8¢ each.......................... .... 4.00
22 yards of small iron chain, @ 2¢ per yard............................ .44
One piece galvanized wire-cloth 22 x 41 inches, with ¼-inch mesh, @ 18¢ per
   square foot, to place over window of storage-room.................... 1.25
4 sheets of No. 24 galvanized iron to cover the furnace, 16 pounds to sheet,
   @ 6¢.. 3.84
Asbestos sheathing, 6 pounds to the 100 feet; one piece of sheathing for
   lower inside furnace-casing, 15 feet ½ inch by 18 inches; one piece of
   sheathing for middle section of inside casing, 15 feet ½ inch by 22
   inches; about 52 square feet @ 15¢ per pound, ${}_{10}^{9}$¢ per square foot.... .47
One piece around radiator on inside casing, 27 feet square, about.......... .25
Inside lining of sheathing under flat top of furnace, 68 inches in diameter,
   with waste, about 5¾ square feet................................ .05
Galvanized-iron washers, 1½ inches in diameter, are used to fasten sheathing
   to the casing by being riveted through; they can be cut from waste.
Asbestos cement fitting to cover flat top inside casing rim, 1 inch thick,
   circle 5 feet 8 inches in diameter, 5 square feet, @ 8½¢ per foot........ .42
Rivets for above work, about.. .10
Sheathing for water-closet hot-air pipe on the first floor, 4 x 12, cut 42 x 34
   inches, about 10 square feet, @ ${}_{10}^{7}$¢ per foot................... .09
Sheathing for 67 feet of 12-inch hot-air pipe; 20 lengths, each 42 x 39
   inches; each length about 11½ square feet, @ 10¢ per length........... 2.00
90 feet 6 inches of 10-inch hot-air pipe to be covered, about 26 lengths, each
   length 42 x 33 inches, about 9 square feet to each length, @ ${}_{10}^{7}$¢ per foot,
   8${}_{10}^{1}$¢ each length, about 2.11
70 feet 4 inches of 4 x 20 inch tin pipe, about 21 lengths, each length 42 x 50
   inches, about 15 square feet in each length, @ ${}_{10}^{9}$¢ per foot, 13½¢ per
   length... 2.84
30 feet 9 inches 4 x 12 inch tin pipe, 16½ lengths, each length 42 x 34
   inches, about 10 square feet, @ ${}_{10}^{9}$¢ per square foot, 9¢ per length, about 1.47
9 feet 8¾ inches tin pipe, 2½ lengths, each length 42 x 29 inches, about 8½
   square feet, @ ${}_{10}^{9}$¢ per square foot, about 8¢ each length............. .20
One tapering joint of tin pipe from 12 to 10 inches in the closet adjoining
   kitchen; two sheets of tin 14 x 20 inches.
Labor putting up furnace, pipes, registers, &c., 11 days for hand and helper;
   mechanic, $3 per day; helper, $2 per day.......................... 55.00

    Total.. $510.86
Add 33⅓ per cent. for profit.. .... 170.28
    Total $681.14

VENTILATION.

The word ventilation is derived from the Latin word *ventilo*, and is thus defined 1, to fan with the wind; 2, to open or expose to the free passage of the air or wind; 3, to cause the air to pass through. The object of ventilation is to insure personal health, domestic cleanliness and pure air in our habitations. Pure air is composed of three gases, viz.: Oxygen, 23; nitrogen, 76; carbonic acid, $\frac{4}{10}$ of 1 volume in every 1000 volumes, and a variable amount of atmospheric vapor, which, though small, is just as necessary for comfort as oxygen is for its life-giving properties. It is a fact that ventilation is just as important as drainage to houses, and that man can no more live in a foul atmosphere than he can survive constantly drinking polluted water. Ventilation is a necessity caused by modern ways and customs. We owe most of our knowledge of it to late discoveries. Although this subject was written upon and its theories discussed and advanced by the celebrated Dr. Desaguliers during the seventeenth century, it is only of late years that a practical and satisfactory application of its principles has been consummated. *Composition of the Air. Importance of Ventilation.*

The one great and grand principle of ventilation is to keep the air in motion. Without this motion there can be no ventilation. If a house or room built moderately air-tight be kept closed for a short time you will find upon opening it a close or musty smell. Rain-water if kept at rest, which is the case in pools and ponds around vacant city lots, becomes stagnant. Flesh in the same manner becomes tainted when there is no longer a blood motion through it. Malaria in certain sections is supposed to be largely due to an absence of motion in the air, allowing it to become impregnated with the poisonous exhalations from stagnant water in ponds and marshes. All of these conditions are detected by the sense of smell. A man can live from eight to sixteen days without water. If water be given without food life can be sustained much longer, as witness the experiment of Dr. Tanner. When deprived of air life is extinct in a few minutes. This undoubtedly shows the value of ventilation. The premises around all houses should be thoroughly drained and kept in a cleanly condition, for there is no sense in changing the already damaged air in the house for a worse article. *The Principle of Ventilation.*

There are two kinds of ventilation, natural and artificial. Cold air is much heavier than warm air, consequently within a certain hight of atmospheric space it obeys gravitation and falls nearest the earth, while the warm air ascends. As the breath leaves the nostrils the whole of it ascends, not only *Methods of Ventilating.*

because it is warmer and lighter, but because the outer air is colder and heavier; this by its expansive force pushes the breath out of the way before the next respiration. If this breathing happens in the open air the carbonic acid gas contained in our exhalations falls to the earth, there to be scattered by the shifting winds through the atmosphere, where it will resume its normal proportions. The weight of the air taken as 1, the weight of carbonic acid gas is 1.52901, showing it to be more than one-half heavier than air. Notwithstanding this when heated its expansive force is greater than air. When the breathing happens in an open hall or crowded room the carbonic acid gas will fall to the floor by the same law; by opening the windows and doors in a proper way and at the proper time we have natural ventilation. The word proper has an extensive significance in this connection. Artificial or forced ventilation is accomplished by means of force and exhaust wind fans, by allowing the air in the exhaust or ventilating flue to come in contact with the radiating-surface of the smoke-pipe, which is carried up through the center of the flue to the top of same, by means of an open fire in the fire-place, and by introducing a current of heated air from the heating apparatus with which the house is supplied into the ventilating-flue. All of these methods, except the open fire produce an upward current in the ventilating-flue. By connecting the flue at the floor-line with the room by means of a register the cold air near the floor is absorbed, exhausted or carried up the flue by means of the upward current produced by the warm air, allowing the hot air from the heating-register to take its place, which produces a constant circulation or motion in the air, which is the true principle of ventilation. The open fire in the fire-place by the draft which is produced in the chimney causes circulation or motion in the air on the same principle, but if a damper is not used in the throat of the chimney to control the draft or outflow of the products of combustion the exhaust or outflow of the air from the room cannot be controlled.

Air-Supply Required.   It is a physiological question of vital importance to determine how much space and the necessary amount per head per hour are required to keep the air pure and fresh. When one takes a long breath the air passes through the throat and windpipe, then through about 1700 bronchial tubes into the air-cells of the lungs. These air-cells have been counted and measured. There are about 6,000,000 of them and each one of them measures about 1-75 inch in diameter, the air-cells holding together 20 cubic inches of air. Each individual breaths 18 times a minute, 360 cubic inches, therefore each individual damages 16 cubic feet per hour. That which we breath in contains 4·10 volume in 1000 volumes and that which we expel contains 4 volumes per 1000 of carbonic acid. It has been discovered by careful experiment that taking the relative diffusion of the carbonic acid with the other gases it is found that 2000 feet of pure air must be furnished each person per hour. In the designing of buildings, public and private, the architect should see that provisions are made for supplying the above amount.

For perfect ventilation to have an outflow or exhaust to carry off the vitiated air of the room; this outflow should be about 11 square inches per head. It is an easy matter to determine if a flue is an outflow or inflet by means of a lighted candle. An ordinary chimney, for example, 12 x 12 inches, with a medium upward draft established will discharge about 28,000 feet of air from an apartment, which if replaced by pure air will furnish a sufficient amount for 14 persons. *Size of Exhaust Flue.*

The object of this essay is to consider the best and most effectual plan for ventilating the dwelling the heating of which I have just finished. The parlor, library, dining-room and the rooms over them on the second floor have open fireplaces with movable grates. The contracted throats of fireplaces will be closed by dampers easy of adjustment. This provides the way of cooling off the rooms if overheated in winter and admits of a positive circulation of air in summer. In the early spring and fall, when it is necessary to take the chilly air from the room without being compelled to operate the furnace, very often sickness requires a warm temperature in the bed-chamber and it is very convenient to have the ready means at hand for furnishing it. For these reasons I propose to let the fire-places be used for what they were intended—namely, an open fire. *First Floor.* *Fire-Places.*

The kitchen chimney should be built with a ventilating-flue 12 x 12 inches and a smoke-flue 9 x 9 inches with a 4-inch division-wall between them; both flues should be smoothly plastered and the brick laid in rich mortar. The outside dimensions of this chimney would be 33 x 20 inches. The ventilating-flue is sufficiently large to carry off the vapors from the laundry, the aroma of cooking from kitchen and the foul air from the rooms above kitchen. In the laundry near the ceiling I propose to use a 6 x 10 inch register connected with the ventilating-flue to carry off the vapors from washing, &c. In the kitchen I propose to have a canopy placed over the range and in the ventilating-flue directly under the canopy at its highest point an 8 x 12 inch register to carry off the aroma of cooking and to cool the kitchen in summer. *Ventilating-Flue.*

Second Floor. The bed-chamber over kitchen to be connected to the ventilating-flue by a 6 x 10 inch register at the floor-line; this
outflow will be sufficient for about four persons. It is also
my intention on this floor to connect the bath-room with this
flue by placing a 6 x 10 register in a tin register-box with a
3 x 12 inch tin pipe and circle elbow swung from the top of
register-box, the register to be placed at the ceiling-line and
the pipe run between the joists in a line with the ventilating
flue of the kitchen chimney. This is to carry off the foul,
heated air which is so prevalent in bath-rooms having water-
closets.

Attic. On the top floor I intend to connect the attic with the
kitchen ventilating-flue by placing a 6 x 10 inch register at
the floor-line. This completes the exhaust or outflows for
the rooms above mentioned.

Around all chimneys are placed trimmers or short joists
which set off from the chimney from 1 to 2 inches, into which
the regular joists which run against the chimney are mitered.
This prevents any wood from coming in contact with the
chimney and lessens the danger from fire. In running a tin
ventilating-pipe parallel with the joists the same as from the
bath-room above mentioned, whether it be at the ceiling or
floor line, in passing through this trimmer you will have to
cut out enough of same to let the pipe lay flush with the
joist. This should always be cut from the top of the trimmer
next the floor and should never be over 3 inches; this will not
weaken it.

Chimney
Ventilating-
Flues. In regard to the chimneys in the parlor, library and dining-room, they will each require the addition of a ventilating-flue 12 x 12 inches. In the parlor chimney this flue
should be built on the south side, and the fire-places in the
parlor and chamber over parlor should be gathered to the
north side of the chimney, dispensing with the short flue
leading from the childrens' play-room ; this will admit of the
building of the ventilating-flue without increasing the size of
chimney. In the library chimney a ventilating-flue 12 x 12
inches should be built on the east side of chimney, and the
fire-places in the library and chamber over library gathered as
on original plans. This can also be accomplished without increasing the size of chimney. In the dining-room a ventilating-flue 12 x 12 inches should be built on the east side and
the smoke-flue running to cellar carried upon the west side
of chimney. The fireplace-flues in dining-room and chamber over dining-room to be carried up between the ventilating-flue on the east and the smoke-flue on the west. The
chimney on the original plan being contracted as it rises
through the roof, it will be necessary to build the chimney

the same size at the top as at its foundation. This will admit of all the necessary flues being built in the chimney. It is generally advised by authors on ventilation to use the smoke-pipe from heating or cooking apparatus for ventilation, by running it through the center of ventilating-flue to its top, for the waste heat radiated through the smoke-pipe to the sur-rounding air in the flue produces an upward or exhaust cur-rent. Through registers placed at the floor-line and con-nected with the flue the foul air is carried from the room by this exhaust current, making room for entrance of pure air and producing a circulation or motion in the air. This is a good plan for schools and other large buildings whose style of architecture will admit of large and thick chimneys, but in dwellings where the walls of the chimney are but 4 inches thick this plan is dangerous. The smoke-pipe put up on this plan requires to be braced at different places in the chimney to keep it in the center of the flue and to support its weight. If made of light material it does not last long and its renewal causes considerable annoyance and very often injury to the chimney. If made heavy of cast-iron or terra-cotta pipe it takes up a large space, and its great weight is a con-stant menace to overstraining the chimney, causing the sup-ports to give way, the smoke-pipe to settle and often separate, which, no doubt, has been the cause of a number of fires at-tributed to defective flues.

I propose in this dwelling to use a system which has the merit of being cheap, easy of control and positively safe. In the ventilating-flue of the parlor, library and dining-room chimneys I propose to introduce hot air from the furnace by means of hot-air pipes run into ventilating-flues of said rooms. I propose to use a 5-inch tin hot-air pipe to connect from fur-nace to ventilating-flues of the library and dining-room, and to parlor ventilating-flue a 4-inch pipe, as the distance from the furnace is very short. This will afford all the heat re-quired for an upward current. The actual amount of fuel re-quired to furnish heat to fill these pipes will be very small, as the combined area of the three pipes is but a small fraction over the area of an 8-inch pipe; besides it would seldom be necessary to have these carry their full capacity of hot air. The pipes should be well wrapped with asbestos sheathing to prevent loss of heat in the transit from the furnace to the flue. *System of Ventilation Proposed.*

The tops of all the ventilating flues should be covered with a stone flagging or heavy metallic plates elevated upon piers 12 inches high built at the corners of the flues. This is to prevent a down-draft when the wind comes from certain quarters. This mode of covering the chimney is adapted to all styles of architecture. *Ventilating-Flue Tops.*

Ventilating-Registers.   In the parlor I propose to place at the floor-line an 8 x 12 inch ventilating-register, which will be large enough for six or seven people. As the parlor will not be continually occupied this will afford ventilation for a larger number. In the chamber over parlor a 6 x 10 ventilating-register. This is to be set n a tin box at the floor-line at the base of closet situated on the south side of the parlor chimney and connected with the ventilating-flue by a 3 x 12 inch tin pipe running from register-box to flue. This pipe will rest upon the floor of closet. This will furnish ventilation for four people. In the children's play-room I propose to use an 8 x 10 ventilating-register at the floor-line connected with ventilating-flue. My reason for using such a large size in comparison with the chamber below is that it will be used principally for the children, who are short of stature, consequently their breathing is taken from the lower strata of air in the room, and it is doubly important that there should be a large exhaust for foul air, which of course can be regulated at will by opening or closing the register in the ventilating-flue. It is well known among physicians that children are more susceptible to colds and other diseases on account of their shortness of stature, which compels them to breath the impure air which lies nearest the floor. I propose to connect the chamber over the dining-room on the top floor by placing a 6 x 10 ventilating-register and the register-box at the floor-line in the northeast corner of room, on a line with the parlor ventilating-flue, connecting a 3 x 12 inch tin pipe with an elbow to bottom of the register-box and running pipe between the joists to ventilating-flue, making the connections the same as for the bath-room on the floor below. This completes the exhaust or outflow for the rooms mentioned above. I propose to use in the library an 8 x 12 inch ventilating-register connected with the ventilating-flue at the floor-line; this will furnish an exhaust or outflow for the same number of people as the parlor. In the chamber over the library a 6 x 10 inch ventilating-register placed at the floor-line and connected with ventilating-flue, which will furnish an outflow for four people. In the billiard-room on the top floor over the library a register 8 x 10 inches placed at the floor-line and connected with the ventilating-flue ; this will furnish an exhaust or outflow for five or six people. I also propose to place a 6 x 10 inch ventilating-register at the ceiling-line connected with the ventilating-flue. It is usual for those occupying the billiard-room to smoke, and when such is the case the upper register can be opened and these vapors quickly carried from the room. In the dining-room I propose to use an 8 x 12 inch ventilating-register connected to ventilating-flue at the floor-line, with the same average for outflow as in the parlor. In the chamber over dining-room a 6 x 10 inch ventilating-register connected with ventilating-flue at the floor-line. This outflow is sufficient for four people. The chamber over dining-room on top floor, as before mentioned, is connected with parlor ventilating-flue. This completes all the apartments in the dwelling except the sewing-room and the water-closet on the first floor. As for the sewing-room, its size is so small that it cannot be well occupied by more than two persons at a time and so situated that by opening a door direct communication so made with the hall, which will furnish all the ventilation necessary. As will be observed, each room has individual ventilation.

Each register placed in the ventilating-flue will require a tube or box to be built in the flue when it is erected. It *Ventilating-Register Boxes.* should be made of galvanized iron and should be made of such a size that the register will fit it tightly. This tube should be 6 inches long, with a ½-inch edge laid off at one end and at right angles to the box a strip 1 inch wide to rivet to tube at right angles to it. This strip should be placed on the four sides of the tube 4 inches from the ½-inch edges. This is the width of a brick, and when the tubes are built in the ½-inch edge will bind against the inside of the flue and the inch strip against the outside of the flue. This will prevent the box or tube from shifting or becoming loose after being built in. The tube is made 6 inches long for the following reasons: Width of brick in the flue, 4 inches; thickness of plastering, ¾ inch; thickness of baseboard, ⅝ inch; edge, to bind against inside of flue, ½ inch; total, 6 inches. These tubes should be built in above the joist of such a hight as to allow for thickness of floor, hight of base-board, and allow for width of flange of register. Where the rooms are to be connected to the ventilating-flue by tin pipe run between the joists the tubes should be made to fit the tin pipe neatly and finished as described for the other tubes. All these tubes or boxes should be built in the flue when it is being erected. The tin register-boxes connected to ventilating-flues should be placed in the studding and the pipe run between the joists before the dwelling is lathed.

The ventilating-registers should be set with their widest part in the width of the flue, as this will bring the register-openings nearer the floor, making them much more effective in carrying off the foul air from the room. All of the registers at the floor-line can be of the vertical-wheel pattern, and black-japanned, as they are not easily discolored. The registers at the ceiling-line should be white-japanned, of self-indicating pattern. They should have chain attachments and wood handles, to permit of their being operated, instead of the cord and tassle arrangement, which is apt to collect dust, get moth-eaten and after a short time present an unsightly appearance.

The general rules given for making pipes, regulator, boxes, &c., in furnace heating can be followed in making up materials for ventilating work. All work should be done in a substantial manner, the best quality of IX bright charcoal tin used, as ventilating work after being one placed in position cannot be easily gotten at.

Fire-Place Throats. In describing the construction of the fire-place I make mention of a cast chimney-throat with a damper and handle easy of adjustment and operation. This is a valuable improvement in fire-place construction and is meeting with much favor at the hands of architects and builders. It is smooth inside, has no dead air-spaces, no projections to prevent the smoke from going direct to the chimney. It is perfectly tight, avoiding danger from fire, has a perfect working damper, requires no special skill in setting and is a valuable adjunct in controlling the escape of smoke, heat and air up the chimney-flue.

VENTILATION ESTIMATE.

I will now consider cost of material and work for ventilating the dwelling as before described.

54 feet 5-inch tin pipe from furnace, 54 sheets 14 x 20 inches, @ 8¢	4.32
3 feet 4-inch tin pipe from furnace, 2 sheets 14 x 20 inches, @ 8¢	.16
23 feet of 3 x 12 inch tin pipe, 17 sheets 20 x 28 inches, @ 16¢	3.91
Two circle elbows 3 x 12 inches, 1 sheet tin 20 x 28 inches, @ 16¢	.16
Two 6 x 10 register-boxes 5¼ deep, 1½ sheets 20 x 28 inches, @ 16¢	.24
Labor for making above work, 1½ days, @ $3	4.50
Galvanized sheet-iron tubes or boxes for flues: seven tubes for 6 x 10 inch register-boxes, size 6 x 33 inches; three tubes for 8 x 12 inch register-boxes, size 6 x 40¾ inches; two tubes for 8 x 10 inch register-boxes, size 6 x 36¾ inches	13.29
Two tubes for 3 x 12 inch ventilating-pipe, 6 x 30¾ inches; one tube for 5 x 8 inch ventilating-pipe, for the closet in chamber over parlor, 6 x 26¾ inches, with binding strips, about 26 pounds, @ 6¢	1.56
Labor for above work, three-quarters of a day, @ $3	2.25
Two 5-inch spring dampers for hot-air pipe at the furnace, @ 15¢ each	.30
One 4-inch spring damper for hot-air pipe at the furnace, @ 15¢	.15
Four 8 x 12 registers, @ 84¢ each	3.36
Two 8 x 10 registers, @ 75¢ each	1.50
Nine 6 x 10 registers, @ 63¢ each	5.67
One joint 5 x 8 inch tin-pipe, 1 sheet 20 x 28 inches, @ 16¢	.16
One 6 x 10 register-box 27 inches long, 2½ sheets 20 x 28 inches, @ 16¢	.40
Labor for this work, 2 hours' time, @ $3 per day	.60
Solder for above work, about 1 pound, @ 20¢	.20
One pound of 6d nails, @ 4¢	.04
Wire	.05
Small tinned chain for ceiling-registers, 9 feet each, @ 5¢ per yard	.30
Four wood handles, @ 5¢ each	20
One 5½ galvanized collar for the cellar wall, for the 5-inch ventilating, to run through, 15 inches long, 1½ pounds, @ 6¢	.09
Two 5-inch galvanized iron collars at furnace, cut 8 inches long, about ⅝ pound each, @ 6¢	.10
One 4-inch galvanized-iron collar at furnace, cut 8 inches long, less than ⅝ pound, @ 6¢	.04
Labor for collars, about 1½ hours, @ $3 per day	.45
Asbestos sheathing for three hot-air ventilating-pipes in the cellar landing from the furnace: for 5-inch tin pipe, 15½ lengths each 17 x 42 inches, about 5 square feet, @ $\frac{9}{16}$¢ per square foot, 4½¢ each length, about	.70
For 4-inch tin pipe in the cellar, one length 14 x 42 inches, about 4 square feet, @ $\frac{9}{16}$¢ per square foot, about	.04
Labor for putting above work in position in dwelling, two and one-half days hand and helper, @ mechanic $3 per day, helper $2 per day	12.50
Six improved cast chimney-throats for open fire-places, with dampers and handles, @ $3 each	18.00
	$61.95
Add 33⅓ per cent. for profit	20.65
	$82.60

The waste from pipe can be used for bands to fasten ventilating-pipes, &c. The registers on the plans marked V denote ventilating-registers, and I have placed them at the points desired, the same as if the flues were upon the plans.

HOT-AIR SYSTEM.*

BY E. E. DUNNING.

SPECIFICATION.

In locating furnace as shown, Fig. 1, it should be placed Location of Furnace. in the most central position, but a very little to the northwest of rooms to be warmed, as the prevailing winds are from that direction. In locating the registers and stand-pipes as here shown, I have aimed to have the pipes as short as can be and as near of equal length as possible. I have indicated on pipes on basement plan, Fig. 1, their size, also by the figures 2 and 3 the story which the stand-pipes lead to, and on first-floor plan have indicated size of stand-pipes in wall, and have governed Size of Hot-Air Pipes. size of pipes and registers by size of rooms and their location. Rooms on first floor require larger pipes than those in second story, and the rooms in attic still smaller pipes for the same amount of air to be heated, as the longer the perpendicular pipe the more heat will pass through the same sized pipes, and, furthermore, rooms with lower ceiling are easier to warm.

In placing the furnace the cellar should be excavated Setting Furnace. where the furnace is to stand 14 inches deeper than the floor-level. First put in a grout and cement bottom under the place where the furnace is to stand, then build of brick a circular wall 14 inches high, which will bring the top level with the floor of the basement. Place on this foundation a furnace Furnace. with casing 54 inches in diameter and a fire-pot 30 inches in diameter, the furnace to consist of cast-iron ash-pit with revolving bar-grates, operated with a crank. On the cast-iron fire-pot I recommend that there be a cast body and steel-plate dome, with steel-plate radiator around it. The dome gives sufficient space over coals for combustion and a radiator around the same adds additional radiating-surface and utilizes more of the heat passing into the chimney.

The smoke-pipe of galvanized-iron to be 8 inches in di- Pipes. ameter, with ventilator and damper. All warm-air pipes to be made of bright cross tin, those in partitions double, with ¼ inch space between, well braced and kept apart. All pipes

* From *The Metal Worker*, January 11, 1890. Copyrighted, 1890, by David Williams.

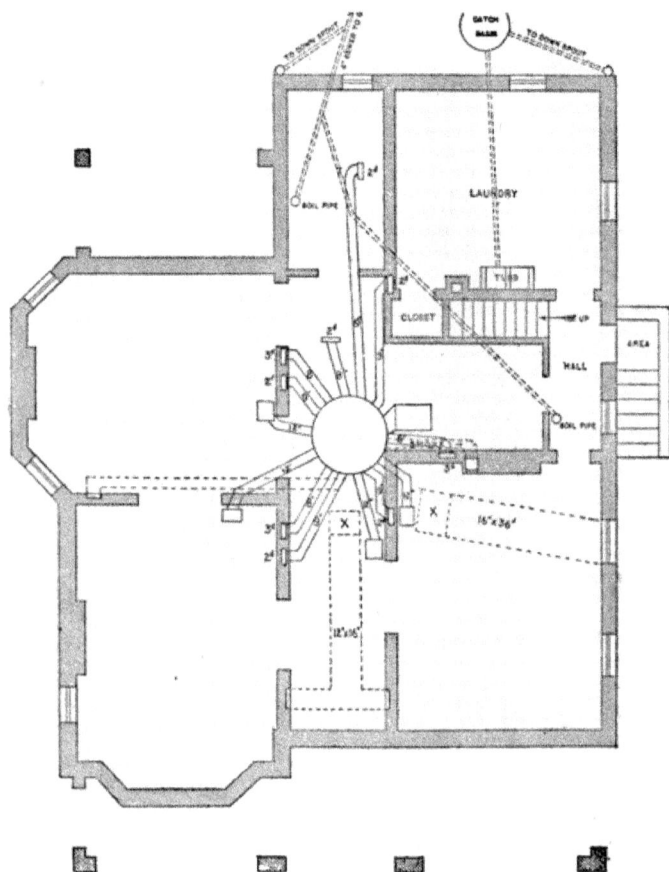

Fig. 1.—Cellar Plan.—Scale, 1-12 Inch to the Foot.

FLOOR PLANS ACCOMPANYING ESSAY OF E. E. DUNNING.

should have as much elevation as possible and single cellar-pipes should be kept 4 inches from wood-work. Where the register-boxes are cover all wood-work around them with bright tin. All tin pipes to have dampers close to furnace, so that heat can be regulated to any desired temperature in any room or entirely closed off if necessary. All elbows should be round, made of three or more pieces and joints double-seamed.

As the prevailing wind is northwest, the cold-air box should be built as shown on plan, Fig. 1, from the north. It should be made of well-seasoned matched lumber, 16 x 36 inside, and in order to take less cellar bottom space can be built on ceiling from window to place marked X (see Fig. 1) and from there down to cellar bottom and thence under furnace. This box should have a tight-fitting damper near window. There should also be a screen over the window and a door in the box at the top of perpendicular part, so that the air can be taken from the top of basement (where the air is more pure than at the bottom) in connection with cold air from hall and rooms at nights or on very cold days and when the wind is in the south, as the outside box will not always furnish cold air when the wind is strong from the south. To further provide for air supply when not using cold air from outside, place a 20 x 26 register in hall as marked, Fig. 2, and build an air-tight box from same 16 x 24 down to under side of furnace. Furthermore, place cold-air registers in parlor and library base-boards, 6 x 14, and build a box along cellar ceiling to place marked X, Fig. 1, and from there down to under side of furnace; also build a box 10 x 12 from cold-air register in dining-room as marked on plan. These boxes should have tight dampers and should be air-tight from register to furnace, made double, with building paper between. It is well and economical of fuel to take air from these places on very cold days and at night.

Build furnace-flues of at least 80 square inches sectional area and with no opening from top to bottom except place for ash-pan, which should always have pan in place, as holes in flues hinder the draft. The mantle-flues should be about same size and have damper over grate to regulate draft. All flues should be of same size their entire length and straight and smooth.

The open fire-places will provide sufficient ventilation in the rooms where they are when taking air from outside. For bath-room, sewing-room and northwest chamber on second floor, Fig. 3, place a 6 x 8 register in base-board and run a 3 x 8 inch pipe through partitions and into kitchen flue in attic, as shown in Fig. 4. This kitchen flue should be

Cold-Air Box.

Cold-Air Pipes.

Flues.

Ventilation.

West

PANTRY
7'0" x 4'

PIAZZA

POTS &
KETTLES
6' x 7'6"

PRESERVES
6' x 6'6"

KITCHEN
18' x 16'4"

PANTRY

5½'x12'

DOWN

PORCH

3'x10'

3'x10'
5½'x12'

DINING ROOM
18' x 18'6"

12"x15'

2½'x28"

TOILET

3'x10'

12"x15"

6"x14"CAR

South

12"x15"

5'x10'
5½'x12'

5½'x12'

12"x15"

North

HALL
7'0"WIDE

12"x15"

PARLOR
18'4"x19'

LIBRARY
18' x 18'

6"x14"
CAR

VESTIBULE
7'0" x 9'

6"x14"
CAR

PIAZZA

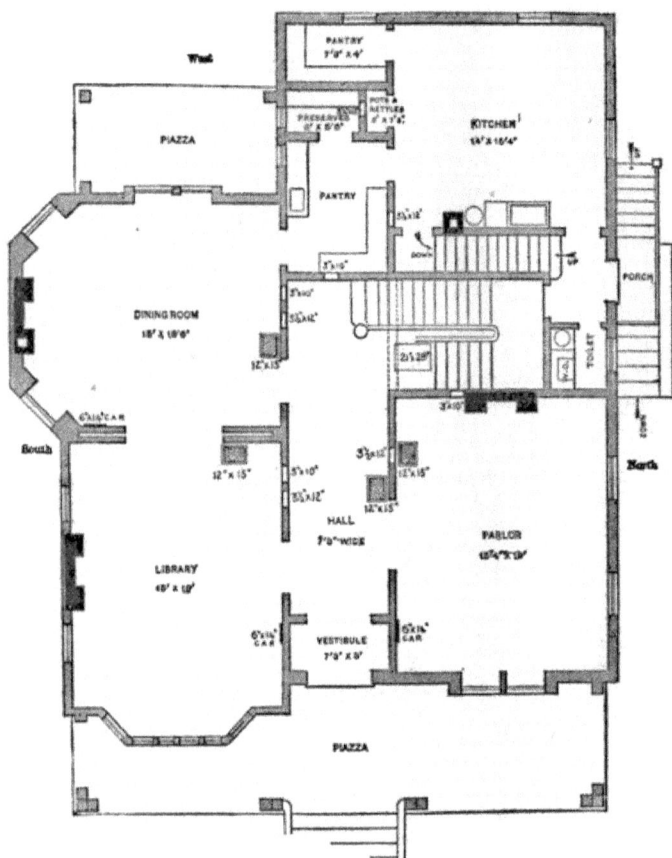

Fig. 2.—First-Floor Plan.—Scale, 1-12 Inch to the Foot.

FLOOR PLANS ACCOMPANYING ESSAY OF E. E. DUNNING.

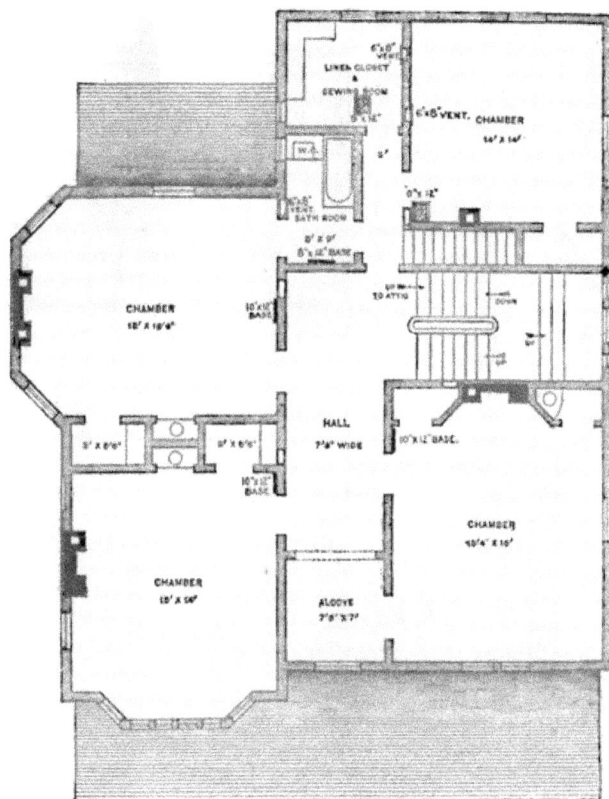

Fig. 3.—Second-Floor Plan.—Scale, 1-12 Inch to the Foot.

FLOOR PLANS ACCOMPANYING ESSAY OF E. E. DUNNING.

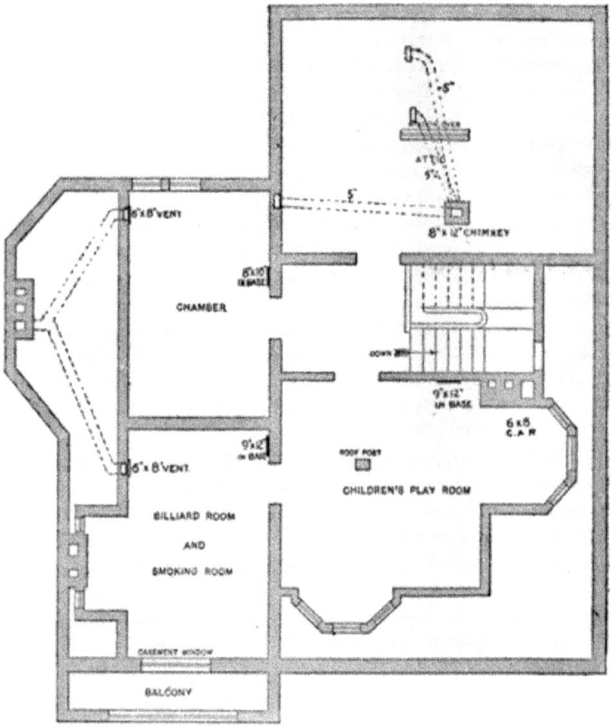

Fig. 4.—Attic Plan.—Scale, 1-12 Inch to the Foot.

FLOOR PLANS ACCOMPANYING ESSAY OF E. E. DUNNING.

8 x 12 inch in order to be large enough to carry off this cold
air and ·not interfere with draft of kitchen range too much.
For the ventilation of the three rooms in attic place a 6 x 8
inch register in base-board and run a 5-inch round tin pipe
from billiard-room and chamber to the south chimney flue,
Fig. 4, which is not used with a fire-place. For children's
play-room place a 6 x 8 inch register above base-board in
grate-flue, as shown in Fig. 4.

I will do the above work as herein planned and specified
for the sum of $358.25, as per estimate.

Cost.

ESTIMATE.

	Cost
1 54-inch furnace	$125.00
250 bricks, 1.80 ; 2 bushels cement, 1.00 ; 1 barrel mortar, 1.00	3.80
1 day's mason and tender work	6.00
1000 feet matched fencing	15.00
400 feet 12-inch lumber, 5.00; Nails for same, 50	5.50
10 days carpenter work	25.00
156 feet stand pipe, @ 50, 78.00; 21 feet 12-inch pipe, @ 30, 6.30	84.30
33 feet 9-inch pipe, @ 25, 8.25 ; 20 feet 8-inch pipe, @ 22, 4.40	12.65
74 feet ventilating-pipe, @ 15	11.10
8 feet 8-inch galvanized iron pipe, @ 30	2.40
3 angle 12-inch elbows, @ 40, 1.20; 4 square 12-inch elbows, @ 55, 2.20	3.40
6 angle 9-inch elbows, @ 35, 2.10; 3 square 9-inch elbows, @ 45, 1.35	3.45
2 angle 8-inch elbows, @ 30	.60
1 square 8-inch elbow	.40
3 galvanized smoke-pipe elbows, @ 50	1.50
13 galvanized collars, @ 20	2.60
13 tin collars, @ 12, 1.56; 14 dampers, @ 20, 2.80	4.36
6 5-inch ventilating elbows, @ 25	1.50
6 6 x 8 ventilating boxes and collars, @ 40	2.40
4 12 x 15 floor register boxes, @ 50	2.00
2 9 x 12 floor register boxes, @ 45	.90
3 10 x 12 base register boxes, @ 45	1.35
2 9 x 12 base register boxes, @ 45	.90
1 8 x 12 base register box, @ 42	.42
1 8 x 10 base register box	.40
4 12 x 15 registers, @ 6.00	$24.00
4 12 x 15 borders, @ 2.40	9.60
3 10 x 12 registers, @ 3.60	10.80
4 9 x 12 registers, @ 3.30	13.20
2 9 x 12 borders, @ 1.50	3.00
1 8 x 12 register, @ 2.80	2.80
1 8 x 10 register, @ 2.50	2.50
6 6 x 8 registers, @ 1.90	11.40
1 20 x 26 register, @ 17.00	17.00
3 6 x 14 registers, @ 3.00	9.00
	$103.30
Sixty per cent. off	61.98

41.32

$358.25

HOT-AIR SYSTEM.*

BY E. M. ROSS.

SPECIFICATION.

Of the three great systems of heating to-day, Steam, Hot-Water and Warm-Air, I have chosen the latter (Warm-Air), not on account of its simplicity or matter of cost, but from pure practical knowledge, knowing it to give better air, less trouble and perfection in heating.

The first and most essential point is to buy a furnace *Furnace.* that is large enough to warm your building, without forcing, and have a reserve from which to draw, as our severe climate so often demands. It is also much better economy, as no more fuel is consumed, the radiating surface is larger, the air is better and the furnace will last longer. The majority of furnaces that are manufactured to-day are too near direct-draft, or in other words the combustion passes directly from the dome or combustion chamber into the smoke pipe, which construction needs no comment, for we all are aware that it draws too heavily on our coal bins. Therefore, one of the first steps of economy is to invest in a furnace that holds the fire smoke and other combustions in the furnace the longest, and the apparatus that is able to do this work, having its proper flues, is the heater to use.

In my experience of heating I find that the portable *Portable Furnace.* furnace radiates more warm air than the brick-set, owing probably to the brick absorbing and retaining the heat. I also have a preference for the steel dome and radiator, as it generally gives better satisfaction; it is quicker to radiate than cast iron and if properly made is entirely gas and dust proof, and will not crack like many cast iron domes, especially where the iron is poor. A furnace should always be made of the best and heaviest material throughout.

In referring to your plans, I see that you have between *Air Space.* thirty and thirty-five thousand (30,000 and 35,000) cubic feet of space to heat, therefore a furnace that will easily heat forty thousand (40,000) cubic feet is none too large. For example, a heater which has a fire-pot that is twenty-seven (27) inches in diameter, inside measurement, and fifteen or sixteen

---

* From *The Metal Worker*, February 15, 1890. Copyrighted, 1890, by David Williams.

(15 or 16) inches in height, weighing between five and six (500 and 600) hundred lbs., following with a dome twenty-nine or thirty (29 or 30) inches in diameter, and thirty (30) inches high. It should also have a proper radiator for retaining the heat as before mentioned, and giving the furnace all the radiating surface that it is possible to heat.

Ventilation. The ventilation is as important as any other part of the work, and perhaps more so, for if the ventilation is not right your work is a failure, and on this subject I enclose you an illustration (Fig. 5) showing the best and only way to remove all the foul air, and get the benefit of all the heat, and the only extra expense is the cost of the furring strips, which run at right angles with the joist, as flooring can be laid as fast on the strips as on the joist. (You might say there is the cost of the cold air shaft and ventilator, but that is not an extra expense, for any house that is properly built will always have its ventilators and shafts.) And by this way of ventilating you not only remove all the foul air but warm all the floors and ceilings, which certainly is economy, and no reasonable mind will deny it, as they can see that it does not take the fuel that it would were everything cold.

Cold-Air Shafts. You will notice by my plans that I have provided for two cold air shafts. One to be built in the center chimney, and one in the south, each being 14 inches by 16, taking all the cold air from each floor, divided up as follows: The parlor, hall, kitchen and pantry should emit at the center shaft, and all rooms directly over. Each floor having a connection with the shaft with an opening which should be 10 x 14 inches. The library and dining-room should emit at the south shaft and all rooms directly over, having the same opening and connections as the other shaft.

Ventilators. You will also notice that in each and every room I undertake to heat I have placed two 4 x 10-inch ventilators, which changes the air throughout the house every few minutes, not only making it a perfect working system, but being one of the most essential points in health.

Location of Furnace. Another very important point is to place the furnace as near the center of the work as possible, dividing up equally all the pipes. What I mean by this, is not to have all the short pipes come on one side of the furnace and the long ones on the other, but place the furnace so you can take a short pipe and then a long one, and in all times where you have the laying out of a job place the registers so that you can make the pipes as short as you can, and always give them all the elevation you have.

Elbows. It is good policy to use as few elbows as your work will admit and never use one at the angle of 90° if you can use

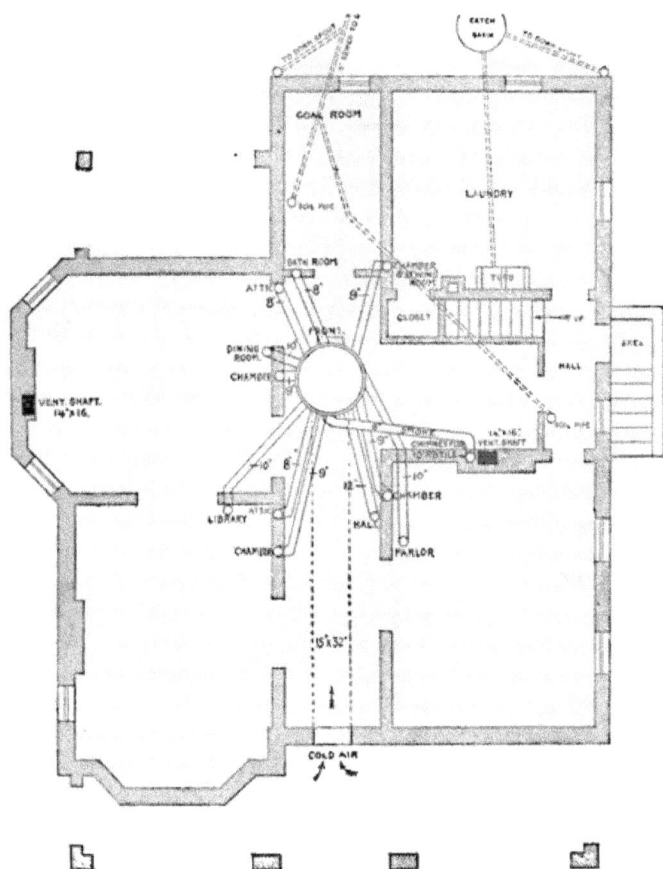

Fig. 1.—Cellar Plan.—Scale, 1-12 Inch to the Foot.

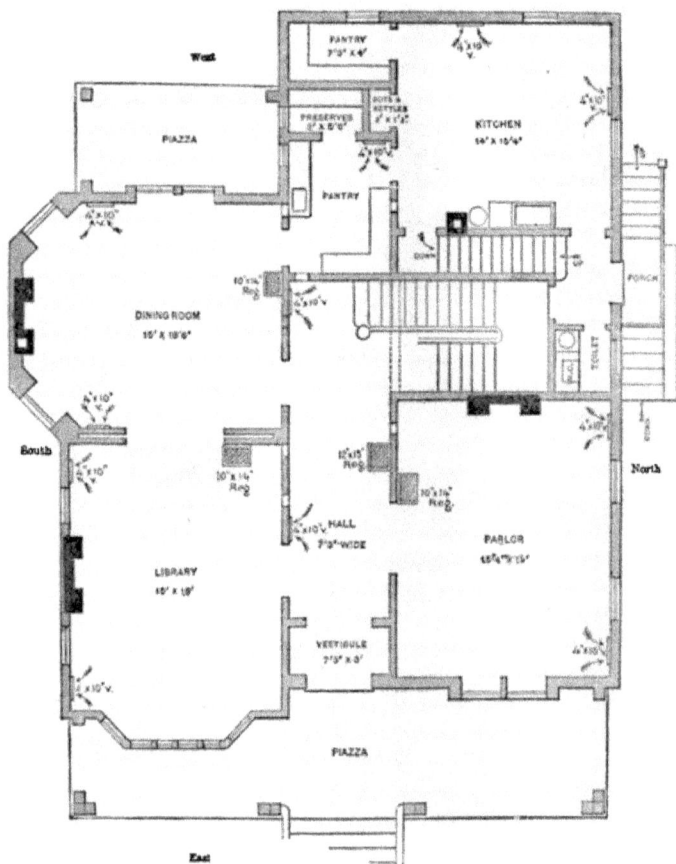

Fig. 2.—First-Floor Plan.—Scale, 1-12 Inch to the Foot.

FLOOR PLANS ACCOMPANYING ESSAY OF E. M. ROSS.

one at 45°, and not at 45° if you can conveniently use 22½°.

Many furnacemen will fasten their furnace collars on to the furnace, never looking to see if they have the proper angle. If they come right, all O. K.; if not, why they start with an elbow or two, which is all wrong. It is just as easy to trim the collar to the right angle and to go to the register-box as quickly as you can get there. A great many times we find traps in hot air work, and the trouble all lies in many elbows. It only needs good judgment to make the work a success. If you are in a hurry to go across the street you do not go until you find a crossing, but go straight to your destination ; and the same application should be made in hot-air work. Go straight from your furnace to your register. *Furnace Collars.*

If you will glance at my plan (Fig. 1) once more you will see that I have located the furnace facing to the west, taking all the cold air from outside from the east. The best way for taking the cold air into the furnace is from the bottom. You should build a pit or well just as large around as your furnace base or bottom will admit, and about fifteen (15) inches deep, connected from outside with a channel that is 32 inches wide and 15 inches deep. It should be built with brick and nicely cemented, having the channel nicely covered over flush with the cellar bottom. There also should be a diaphragm or division in the center of the pit, running from the front to about six (6) inches of the back, so that the cold air will not conflict but rise smoothly into the furnace. *Cold-Air Supply.*

You will note that the dimension of the cold-air channel is a little less than two-thirds (⅔) the capacity of all the hot-air pipes, which amount to 777 square inches of heat, being divided up as follows : *Hot-Air Pipes.*

First Floor.—Hall.—One 12 x 15 floor register and border, connected with the furnace with a twelve (12) inch round pipe. D, dining-room, library and parlor, each with a 10 x 14 floor register and border, connected with the furnace with a 10-inch round pipe. *First Floor.*

It might do well to answer a question here that might arise with some. Why do we use floor register on the first flat ? Generally the rooms are so large that the partition walls will not admit pipes large enough to warm the rooms ; and, again, the flow of heat is much better in the floor. *Floor Registers.*

But in pipes that run to the second and third floors you can use the side wall register just as well, it being on account of the pipes drawing ; and the longer they are the better they work, acting more on the principle of a chimney. *Side-Wall Registers.*

Second Floor.—Northeast chamber, one 9 x 12 side wall register ; southeast chamber, one 9 x 12 side wall register; south chamber, one 9 x 12 side wall register, double header ; hall *Second Floor.*

chamber, one 9 x 12 side wall register, double header; bath room, one 8 x 10 side wall register ; northwestern chamber, one 8 x 12 floor register and border, double header ; back hall, 8 x 12 floor register and border, double header.

Tin Pipes. The second and third stories are connected with the basement with 3 x 10½ inch side wall pipes or pipes inside the partition wall, thoroughly lined with tin for protection, the lining being nailed to the studding, making it, as it were, a pipe within a pipe, connected with the furnace in basement with 8-inch and 9-inch round pipe, designated in Fig. 1.

Attic. Third floor or attic, children's playroom, one 9 x 13 side wall register. Billiard-room, one 9 x 12 side wall register Attic, one 9 x 12 side wall register. Attic chamber, one 9 x 2 side wall register. The attic pipes are both double headers, or two registers off from each pipe, connected in the basement with 8-inch round pipe, making in all eleven pipes off of the furnace and 777 square inches for the heat passing into the different parts of the building.

Chimney Flue. The next very important step is a proper chimney-flue. The flue for this building should be a ten (10) inch round tile, and should be built perfectly straight, commencing in the cellar with a ten (10) inch tee-tile, so that you can have a good clean out underneath your furnace connection.

Grate Flues. Next to the furnace-flue should be the cold-air shaft, and then arrange for your grate-flues, and in the south chimney the cold-air shaft should be between the two grate-flues. The grates should also be provided with dampers, so when they are not in use they could be closed and the ventilation can pass through its proper channels.

Experience is our best teacher, and by adhering to the right principles of heating we always will have satisfaction and a perfect working system besides economy and health.

Regulators. In reference to patent regulators I have but little to say, as I never have seen one that gave the best of satisfaction. A good way to regulate is by chains running over hot-house pulleys from some convenient place on the first floor to the furnace, and by the proper care and study of the furnace you can regulate with this as well as by some high-priced apparatus.

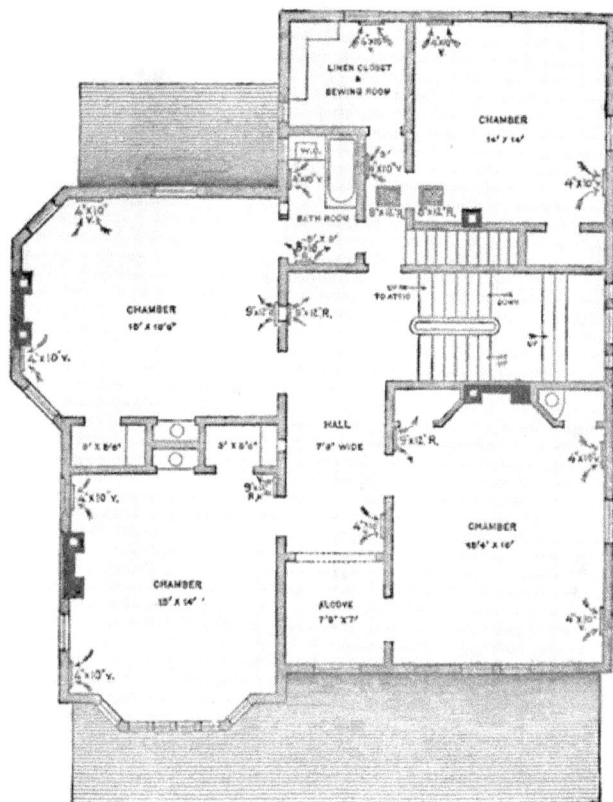

Fig. 3.—Second-Floor Plan.—Scale, 1-12 Inch to the Foot.

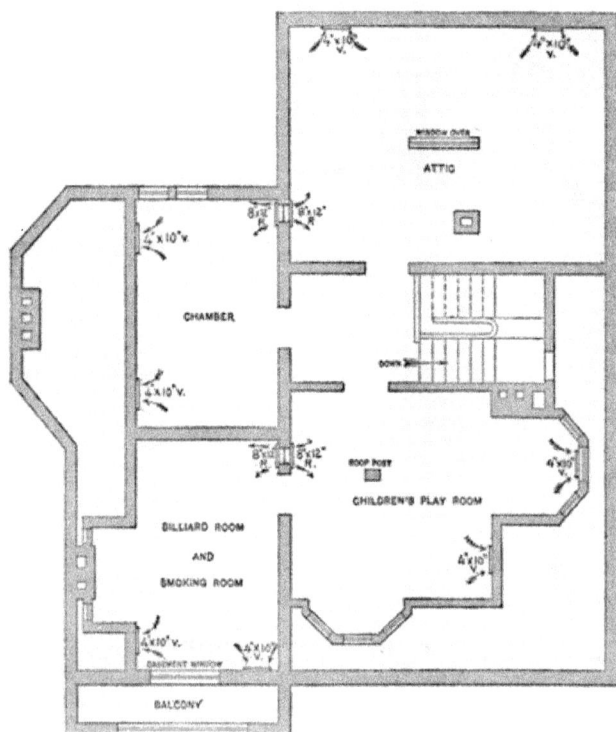

Fig. 4.—Attic Plan.—Scale, 1-12 Inch to the Foot.

FLOOR PLANS ACCOMPANYING ESSAY OF E. M. ROSS.

ESTIMATE.

Basement.

1 large furnace capable of heating 40,000 cubic feet..............	**$150.00**	Furnace.
Building cold-air pit and channel...........................	20.00	
8 feet 12 inches round pipe, @ 30¢.....................	$2.40	Pipe.
28 feet 10 " " @ 25¢.............	7.00	
30 " 9 " " @ 22¢.................	6.60	
20 " 8 " " @ 20¢.......	4.00	
110 feet 3 x 10½ square pipe, @ 30¢ (encased)............	33.00	
12 " 9 inch galvanized smoke-pipe, @ 25¢....	3.00	
		56.00
1 12 inch round elbow, @ 50¢.........................	.50	Elbows.
3 10 " " @ 40¢...................	1.20	
6 9 " " @ 35¢..	2.10	
3 8 " " @ 33¢................	.99	
		4.79
1 12 x 15 register box, 12 inch collar, @ 70¢............	$0.70	Register Boxes.
3 10 x 14 " " 10 " " @ 60¢............	1.80	
		2.50
The side wall boxes are included in with side wall-pipe.		
1 12 inch furnace collar, @ 20¢.......................	$0.20	Furnace Collars.
3 10 " " " @ 18¢.....................	.54	
4 9 " " " @ 16¢.................	.64	
3 8 " " " @ 15¢.....................	.45	
		1.83

Registers.

1 12 x 15 floor register and border in hall	$2.87	First Floor.
1 10 x 14 " " " in dining-room........	2.78	
1 10 x 14 " " " in library.............	2.78	
1 10 x 14 " " " in parlor.............	2.78	
2 4 x 10 ventilators in hall, @ 30¢..	.60	
2 4 x 10 " in parlor, @ 30¢...................	.60	
2 4 x 10 " in library, @ 30¢...................	.60	
2 4 x 10 " in dining-room.....................	.60	
		13.61
1 9 x 12 s. w. register, n. e. chamber..................	$1.65	Second Floor.
1 9 x 12 " " s. e. "	1.65	
1 9 x 12 " " south "	1.65	
1 9 x 12 " " in hall........................	1.65	
1 8 x 10 " " in bathroom............	1 25	
1 8 x 12 floor register and border in n. w. chamber........	2.03	
1 8 x 12 " " " in back hall...........	2.03	

Carried forward..................................$248.73

Brought forward					$248.73
2 4 x 10 ventilator and border in n. e. chamber, @ 30¢				.60	
2 4 x 10 " " in s. e. " @ 30¢				.60	
2 4 x 10 " " in south " @ 30¢				.60	
2 4 x 10 " " in n. w. " @ 30¢				.60	
1 4 x 10 " " in sew room, @ 30¢				.30	
1 4 x 10 " " in bathroom, @ 30¢				.30	
1 4 x 10 " " in back hall, @ 30¢				.30	
					15.21

Third Floor.

1 9 x 12 s. w. register in play-room	$1.65	
1 9 x 12 " " in billiard-room	1.65	
1 9 x 12 " " in attic	1.65	
1 9 x 12 " " in " chamber	1.65	
2 4 x 10 ventilators in play-room, @ 30¢	.60	
2 4 x 10 " in billiard room, @ 30¢	.60	
2 4 x 10 " in attic, @ 30¢	.60	
2 4 x 10 " in " chamber, @ 30¢	.60	
		9.00

Miscellaneous.

11 dampers for hot air pipes, @ 10¢		1.10
7 square pipe collars, @ 15¢		1.05
2 9 inch galvanized elbows for smoke-pipe, @ 50¢		1.00
Tinsmith labor and helper		12.00
Total		$288.09

All work and material to be first-class and executed by competent workmen. The extra work of ventilating and building cold-air shafts will not exceed one hundred dollars ($100.00). I have not allowed for carpenter work, for generally on new houses we make arrangement with the builders, but on this job it would not exceed more than six dollars extra, ($6).

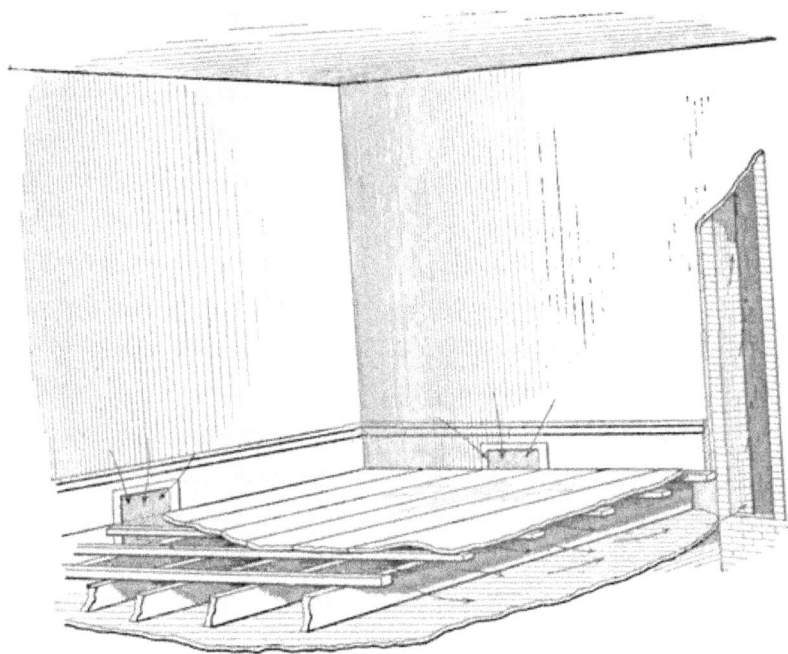

Fig. 5.—System of Ventilation.

HOT-AIR SYSTEM.*

BY A. B. RECK.

SPECIFICATION.

In the annexed tabulated calculation with columns 1–16, Loss of Heat. that explains itself, it is found at the base of column 10 that the total loss in American heat units from windows, walls and ceilings (including loss to cellar through floor of first story) is 96,300 heat units in zero weather. When examining columns 7, 8, 9, in relation to columns 4, 5, 6, it will be found that different rates of heat units per square foot are used in proportion to the different inside temperatures stated in the conditions of the competition, and giving the situation of each room due consideration.

To exemplify, it will be seen that for room No. 1 (see Loss through Ceilings. Fig. 2) are reckoned 9 units per square foot of ceiling, that room having a cold cellar room under it, so that from it heat is passing both down to the cellar and up to No. 11 (see Fig. 3), the temperature at floor in this room being lower than under ceiling in room No. 1. In room No. 11 are reckoned 1½ heat units per square foot of ceiling, because over part of the ceiling is a cold attic room, so that more heat escapes through the ceiling than is received through the floor from room No. 1. In room No. 21 (see Fig. 4) are reckoned 5 units per square foot of ceiling, because all of the ceiling has a cold room over it.

In rooms where the temperature shall be 70° the loss Loss through Windows. through windows is reckoned to be 50 units and through walls about 12 units per square foot of exposed surface. As much is reckoned for the windows to the north and west, though the glass here is double, because it is stated that those sides are more exposed to the winds. The indicated rates of loss of heat units per square foot are what experience has taught will do, but it will be easy to show that they conform very nearly with what can be derived theoretically from such works as Peclet's and Box's treatises on heat.

It is reckoned that the air shall be heated to about 90° Temperature of Heated over the temperature in the rooms. These degrees it shall Air. lose before leaving the rooms through the ventilators at the

* From *The Metal Worker*, March 22, 1890. Copyrighted, 1890, by David Williams.

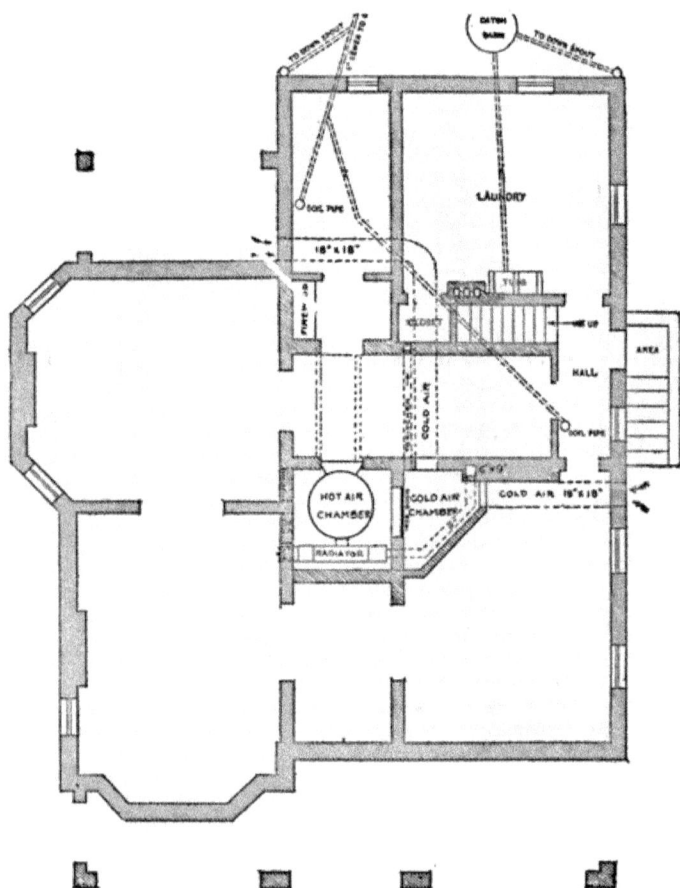

Fig. 1.—Cellar Plan.—Scale, 1-12 Inch to the Foot.

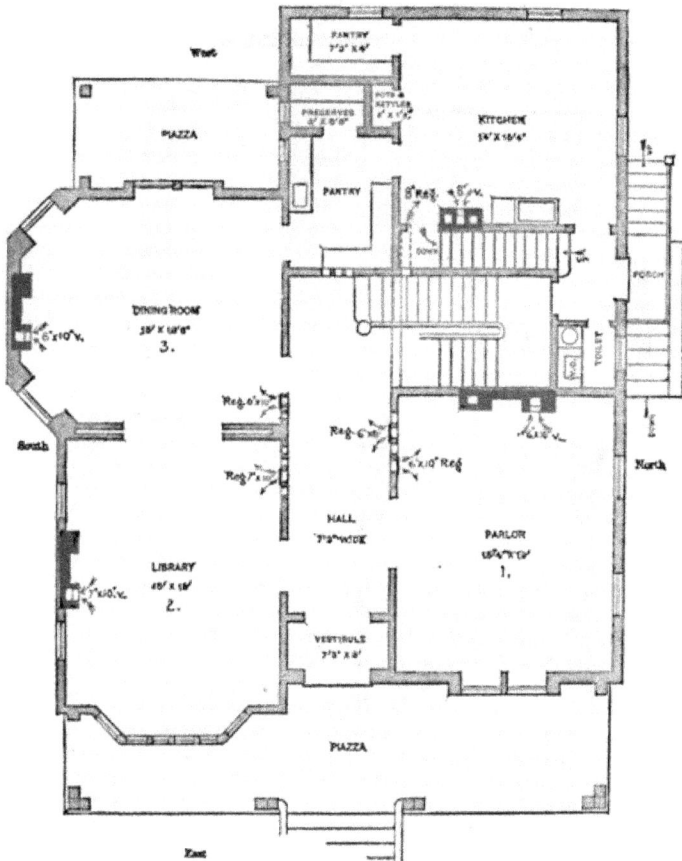

Fig. 2.—First-Floor Plan.—Scale, 1-12 Inch to the Foot.

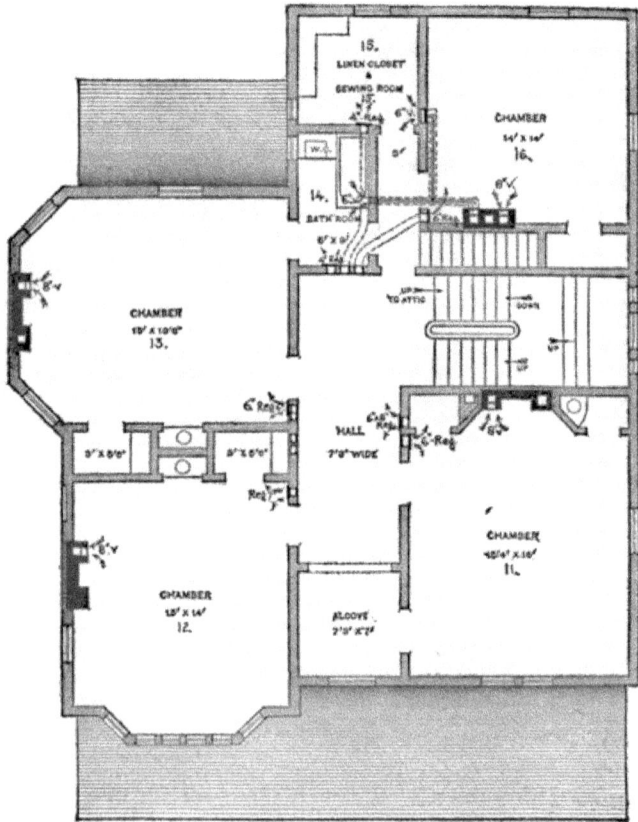

Fig. 3.—Second-Floor Plan.—Scale, 1-12 Inch to the Foot.

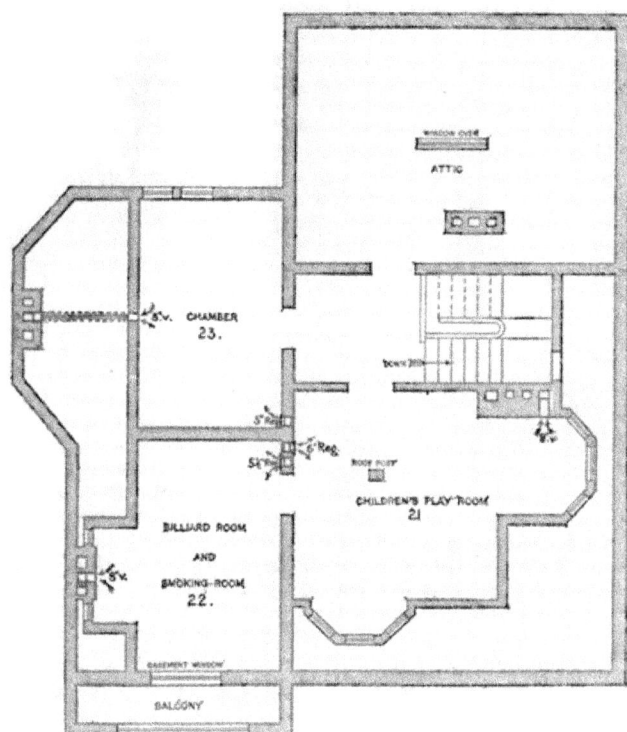

Fig. 4.—Attic Plan.—Scale, 1-12 Inch to the Foot.

floor in the exit flue, and every cubic foot of air shall, while passing through the rooms, give off about 1.8 heat units to the walls, ceilings, &c. When we divide the heat units in column 10 with 1.8, we find how much air should go to each room to heat it, and when we know from experience how fast the air will rise in a hot-air pipe of a given hight—the hight reckoned from 3 feet over cellar floor to hot-air register —we can find how many square inches the area of the flues should be.

Size of Pipes and Registers. When the hot-air registers in the rooms are put in the walls about 7 feet over the floor, the air will in zero weather come in with a velocity of about 6 feet in the first story, 7.5 feet in the second story and 8.5 feet in the third story. The areas of the hot-air pipes will, therefore, be all right if you make them about 6.8, 5.2 and 4.5 square inches for every 1000 cubic feet of air, or 3.7, 2.9 and 2.5 square inches for every 1000 heat units the pipes shall lead to the rooms on each of the three floors. In that manner the areas in column 11 are found, and corresponding hereto the dimensions of the registers and the pipes in columns 12 and 13.

Ventilation. The air leaves the rooms through the chimneys already provided by the architect. Only in three places in the building has it been found necessary to add new exit flues in order that every room may have its own flue. In the exit flues ventilators (see column 15) are put in both near the floors and at the ceilings, the last mentioned to be used when the temperature in the rooms shall be too high, especially in the evening when the lamps are lighted. The exit flues can be used for fire-places in the rooms when the fire-places are provided with dampers in the smoke outlets to the flue. When lighting a fire in the fire-place the damper is opened and the ventilator is shut, and the air will go out of the room through the fire.

Furnace. At the base of column 10 it was found that the hot air must give off 96,300 units of heat per hour to the building, and we have seen that to obtain this $\frac{96,300}{1.8}$ = 53,000 cubic feet of air must be heated to a temperature of 90° over the rooms or about 160° over the outside temperature. To heat 53,000 cubic feet of air 160° takes about 170,000 heat units; that may be given off economically by a good furnace, when it has 1 square foot of heating surface for every 1500 heat units it shall give off. The furnace shall then have 110 square feet of surface (projecting ribs not included) and I propose to distribute this surface in such a manner that 70 square feet is in the furnace proper and 40 square feet in a radiator placed in

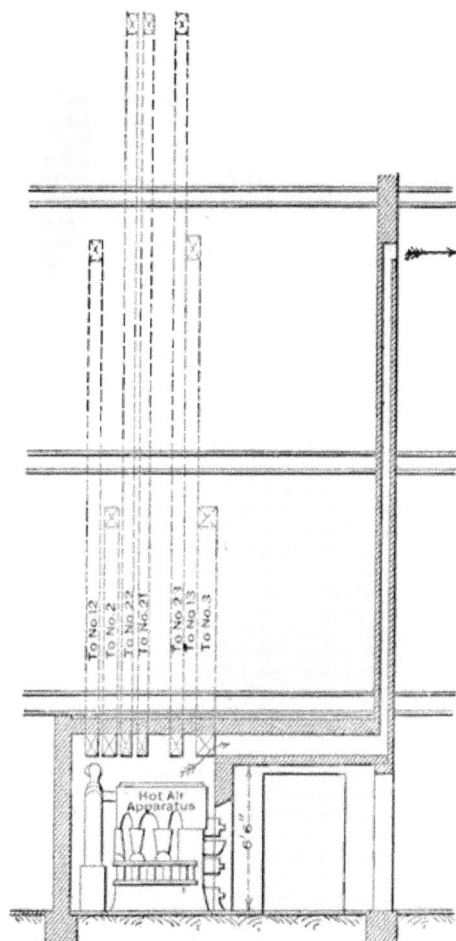

Fig. 5.—Vertical Section through E F, See Fig. 8.

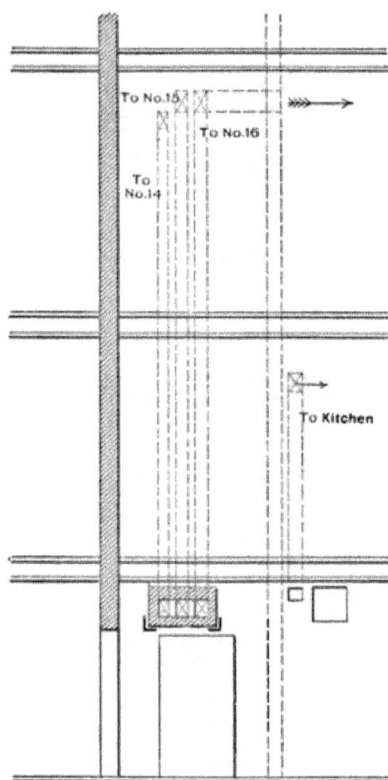

Fig. 6.—Vertical Section through A B, See Fig. 8.

the hot-air chamber behind the furnace (see Figs. 1 and 8). The furnace can be any cast iron, wrought iron or steel furnace with 70 square feet of heating surface of a good make, with tight joints, easy to clean from soot, and with a grate about 28 inches in diameter, easy to shake and clean from clinkers. The furnace shall be provided with a water-pan with a surface of water of 1½ square feet.

I have located the furnace in the middle of the house, where most of the hot-air pipes can go directly up. Only the three pipes to rooms 14, 15 and 16, Fig. 3. are to be led a short distance horizontally before going up. To avoid loss of heat, the air for these three pipes is led in the air-duct of masonry on light iron beams from the hot-air chamber to the vertical air-pipes (see Figs. 5 and 6). The vertical air-pipes are to be made of double tin pipe, with ¼ inch air-space between the two layers of tin to prevent loss of heat and fire risk. *Location of Furnace.*

From the cellar plan, Fig. 1, it will be seen that next to the hot-air chamber is a cold-air chamber to which two fresh-air ducts are led, each 2¼ square feet in area, and beginning from fresh-air inlets covered with galvanized iron wire netting in opposite sides of the house. Each fresh-air duct has a 16 x 24 inch register in the cold-air chamber, so that the supply of fresh air can be taken in calm weather from both the ducts at once, and in windy weather from that duct that gives most air. In this manner good air pressure can always be had in the hot-air chamber and the hot-air pipes from beneath, so that it is to be expected that they will work well to every room, even when the wind is blowing hard against the windows. However, when it should blow too hard it will be possible to heat the whole house by shutting off both fresh-air ducts and all ventilators in the exit flues in rooms. It will then only be necessary to let the doors from the rooms stand a little open to the hall and open the door from hall to cellar and from cellar passage to cold-air chamber. The air in the house will then circulate, going up from hot-air chamber through hot-air pipes to rooms and returning through hall, cellar passage and cold-air chamber. From cold-air chamber the air goes to hot-air chamber under the iron door (see Fig. 7) in the wall between the two chambers, where a space of 4 feet square is left free along the floor. *Cold-Air Ducts.*

To ventilate the kitchen, Fig. 2, an air-duct is taken direct from cold-air chamber to a round register, 8 inch, in the wall of the room. If the owner should wish to use gas or oil for his kitchen stove, and to have his kitchen heated from the furnace during the winter, it will be an easy matter to let the kitchen get hot air from the cold-air pipe. *Kitchen Ventilation.*

The number of cubic feet of air that will pass through each room (see room numbers on plans) is as follows: No. 1, 5900 cubic feet; No. 2, 6500 cubic feet; No. 3, 5600 cubic feet, kitchen, 3000 cubic feet; No. 11, 3500 cubic feet; No. 12, 3900 cubic feet; No. 13, 3500 cubic feet; No. 14, 1700 cubic feet; No. 15, 1700 cubic feet; No. 16, 3100 cubic feet; No. 21, 3800 cubic feet; No. 22, 3100 cubic feet; No. 23, 2700 cubic feet. The dimensions of the hot-air registers and ventilating registers are given on the floor plans.

Table of Exposed Surfaces, Loss in Heat Units, Areas of Pipes, Registers and Ventilators.

1	2	3	4	5	6	7	8	9	10	11	12	13	14	15	16
			Areas.			Loss in heat units.									
Floor.	Room No.	Use of Room.	Windows.	Outside wall.	Ceiling.	Windows.	Outside wall.	Ceiling.	Total heat units.	Free area in hot-air register.	Hot Air Register.	Hot-air flue.	Remarks.	Ventilators in exit flues at floor and ceiling.	Remarks.
1st...	1	Parlor	84	320	290	4,200	3,800	2,600	10,600	40	6 x 10	4½ 9		6 x 10	
1st...	2	Library	110	320	270	5,500	3,800	2,400	11,700	44	7 x 10	4½ 10		7 x 10	
1st...	3	Dining-room	80	300	280	4,000	3,600	2,500	10,100	38	6 x 10	4½ 9		6 x 10	Round Ventilators.
all floors	4	Hall	125	500	330	5,600	5,600	4,000	15,200	28 / 22	6 x 8	4½ 8	1st floor 2d do.	*	
2nd.	11	Chamber	50	320	250	2,300	3,600	500	6,400	19	6 in.	4½ 6		8 in.	
2nd.	12	Chamber	85	270	210	3,800	3,000	300	7,100	21	7 in.	4½ 6		8 in.	
2nd.	13	Chamber	54	300	280	2,400	3,400	600	6,400	19	6 in.	4½ 6		8 in.	
2nd.	14	Bathroom	6	50	45	800	1,800	400	3,000	9	4 in.	4½ 4	Round Registers.	6 in.	
2nd.	15	Sewing-room	20	150	55	1,000	1,600	300	2,900	9	4 in.	4½ 4		6 in.	
2nd.	16	Chamber	36	260	200	1,600	3,000	1,000	5,600	16	6 in.	4½ 5		8 in.	
3rd.	21	Playroom	25	310	350	1,300	3,800	1,800	6,900	17	6 in.	4½ 5		8 in.	
3rd.	22	Billiard room	20	300	190	1,000	3,600	1,000	5,600	14	3½ in	4½ 5		8 in.	
3rd.	23	Chamber	12	300	150	600	3,400	800	4,800	12	5 in.	4½ 4		8 in.	
									96,300						

* No exit flue from hall. Air goes through leaks at doors to rooms.

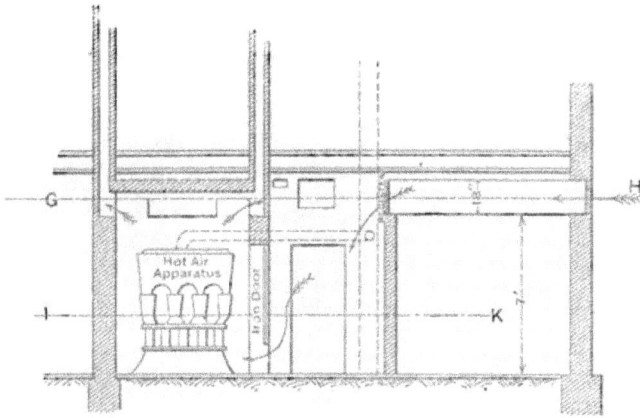

Fig. 7.—Vertical Section through C D, See Fig. 8.

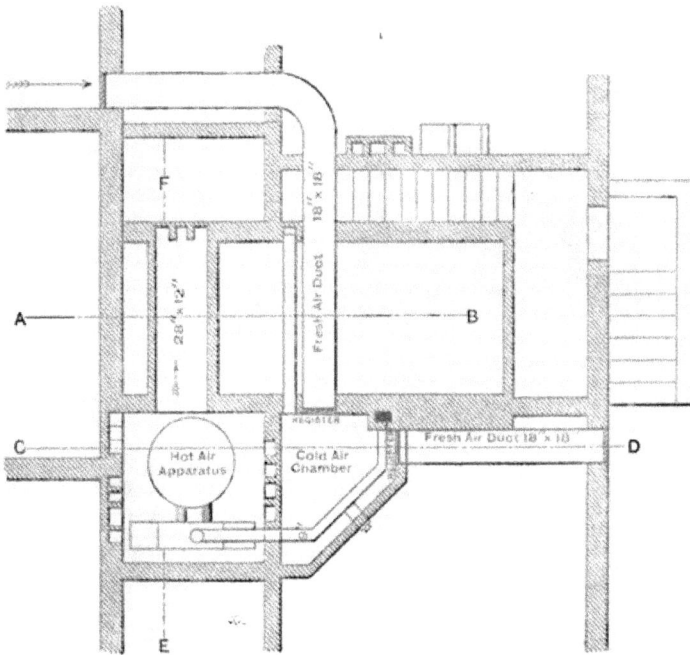

Fig. 8.—Horizontal Section through I K, See Fig. 7.

ESTIMATE.

1 hot-air cast-iron furnace, brick set, 70 square feet heating surface and cast-
iron radiator 40 square feet heating surface, 28-inch grate.................$250.00
Iron bars for hot-air chamber and horizontal hot-air flue over fire-door....... 20.00
9 feet smoke-pipe, 9 inch.. 10.00
Erecting of furnace, including bricks for hot and cold air chambers......... 40.00
35 feet 18 x 18 inch galvanized iron fresh-air duct from outside walls to
cold-air chamber in cellar, @ $1.20.................................... 42.00
2 galvanized wire nettings for fresh-air inlets to duct in outside cellar walls,
@ $1.50.. 3.00
2 pieces 16 x 24 vertical wall registers for fresh-air inlets from ducts to cold-
air chamber in cellar, @ $10.50...................................... 21.00
Iron door, 36 x 48 inches, between cold-air and hot-air chambers.... .... 5.00
Wooden door between cold-air chamber and cellar passage.................. 7.00

Hot-air pipes (including fixing and register boxes).

10 feet 10 in. x 4½ in. to room No. 2, @ 65¢......................... $6.50
20 feet 9 in. x 4½ in. to room Nos. 1, 3, @ 60¢..................... 12.00
32 feet 8 in. x 4½ in. to room No. 4, @ 55¢ 17.50
84 feet 6 in. x 4½ in. to rooms No. 11, 12, 13 and kitchen, @ 45¢...... 38.00

Cold-air pipe to kitchen.

60 feet 5 in. x 4½ in. to rooms No. 16, 21, @ 40¢ 24.00
142 feet 4 in. x 4½ in. to rooms No. 14, 15, 22, 23, @ 35¢ 50.00
                                                    —— 148.00

Registers and ventilators, japanned.

3 pieces 7 in. x 10 in., @ $2.30................................... $ 6.90
6 pieces 6 in. x 10 in., @ $2.10 .... 12.60
1 piece 6 in. x 8 in., @ $1.90 .... 1.90
14 pieces 8-in. round, @ $1.80.................................... 25.20
2 pieces 7-in. round. @ $1.50.................................... 3.00
8 pieces 6-in. round, @ $1.30.................................... 10.40
1 piece 5½-in. round, @ $1.20.................................... 1.20
1 piece 5-in. round, @ $1.20.................................... 1.20
2 pieces 4-in. round, @ $1.10.................................... 2.20
2 wall frames for 7 × 10, @ 60¢.... .. .. 1.20
4 wall frames for 6 × 10, @ 60¢.................................... 2.40
14 wall frames for 8-in. round, @ 60¢............................ 8.40
4 wall frames for 6-in. round, @ 40¢ 1.60
26 cords for 14 hot-air registers and 12 ceiling ventilators, @ 30¢...... 7.80
                                                    —— 86.00

Extra masonry in chimney stacks for 1 smoke flue from furnace and 3 flues for
foul air from chambers.. .... 20.00
Unforeseen expenses, coal and labor for drying out and trial firing.......... 47.00
Salary for consulting engineer for planning the work, controlling the contractor during erection and trials and giving instructions to the servants,
12 per cent. of $700.. 84.00
                                                    ——

       Total....... ....$783.00

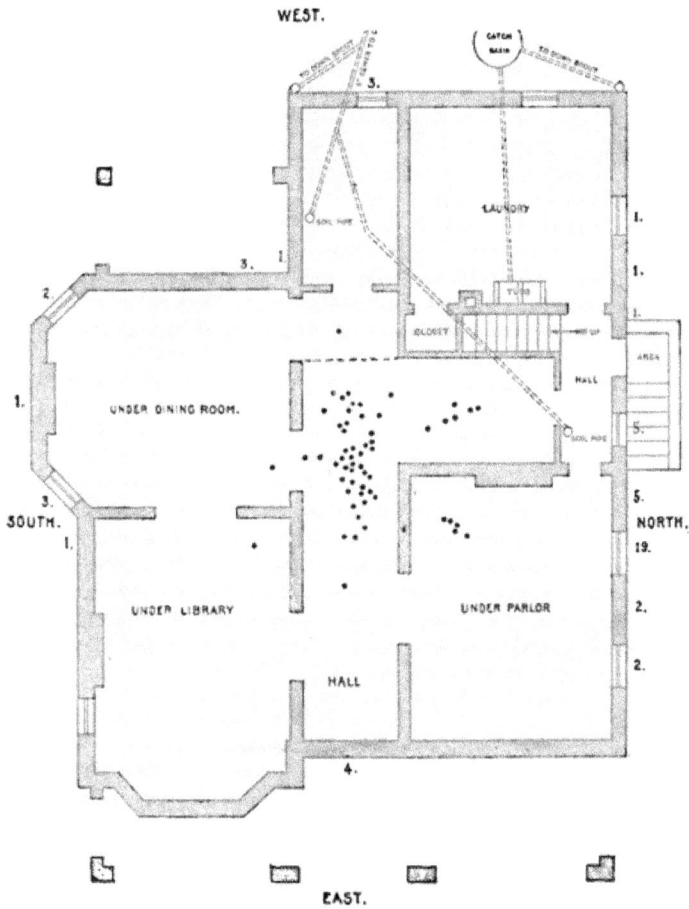

Diagram Showing Location of Furnaces and Cold-Air Openings.

LOCATION OF FURNACES.*

It is probable that in looking over the plans of the house for which an ideal system of heating was to be provided, one of the first points to be considered would be the location of the furnace. In the setting of furnaces it is usual to select a central location, so that the various heating pipes will be as near the same length as circumstances will permit, excepting in some cases where, one side of the house being more exposed than others, it may be necessary to have the pipes leading to that part of the house shorter than others. In the accompanying engraving, which is intended to be a bird's-eye view of the plans submitted as far as the location of the furnace is concerned, the center of each furnace is indicated by a dot. An inspection of the engraving will show that most of the furnaces have been located under the hall and near the center of the house, only one being situated under the library, one under the dining room and five under the parlor. It is seldom that one has an opportunity of observing how others would locate a furnace, as, after it is placed in position in a building, it may only show one person's idea on the subject, but in the present instance an inspection of the cellar plan shows the locations selected by many competitors. Regarding the opening for the cold air duct, as the house to be heated is supposed to be separated from other buildings the competitors have been allowed free choice of location, and have not been restricted, as would have been the case in a city where only one or two sides of a building can be used. The figures on the outside of the plan indicate the number who have selected a certain point from which to derive the cold air supply. Nineteen have indicated that the second window from the front, on the north side of the house, is the proper location, while 36 in all have selected this side of the house, the openings being located as shown by the figures. Only seven have selected the west, six the south and four the east. Regarding the sizes of these openings, there is as great a diversity as in any item connected with the

---

* Reprinted from *The Metal Worker* of December 7, 1889.

various plans submitted. In the following table is shown the sizes of openings or ducts. In some of the plans the location of duct was omitted, and in others the size:

1, 8 x 18.	1, 15 x 36.	1, 20 x 24.
1, 12 x 24.	1, 16 x 22.	3, 20 x 30.
2, 12 x 36.	1, 16 x 30.	2, 20 x 40.
2, 14 round.	3, 16 x 36.	1, 20 x 42.
1, 14 x 20.	1, 16 x 40.	3, 24 x 30.
1, 14 x 22.	1, 17 x 24.	1, 24 x 40.
1, 14 x 36.	1, 18 round.	1, 26 x 40.
1, 15 x 30.	1, 18 x 30.	Average size:
1, 15 x 32.	1, 18 x 42.	15½ x 29½.

Table Showing Size of Cold Air Openings.

In the past there has been more or less correspondence in Furnace Work on the subject of size of the cold air duct in proportion to the area of heating pipes, and in this connection the table is quite interesting, showing as it does the diversity of opinion on the subject.

SIZES OF HEATING PIPES AND REGISTERS.*

Regarding the relative sizes of heating pipes and registers, it will be of interest to summarize and compare the ideas of the contributors to the Hot Air Furnace Competition. Assuming that a 10-inch round pipe is of sufficient capacity to convey the necessary amount of hot air to properly heat a certain room, what size of register should be used so as to produce the best effect ? If a furnaceman is required to estimate from a set of plans at what price the furnace, pipes, registers, &c., can be put in, and, as is apt to be the case, the lowest estimate is to receive the most consideration, it would appear to be unwise to use registers that are larger than necessary, and thereby increase the amount of the bid. Neither should he use those of such small size as to restrict or retard the flow of hot air. It would appear to be quite necessary that in order to make a certain furnace heat a house in an economical manner, the flow of hot air through the registers should not be impeded by a register too small for the pipe.

In Tables I and II are shown, as far as could be obtained from the plans and essays, the sizes of pipes and registers for the parlor, library, dining-room and hall on the first floor and for the four bedrooms on the second floor. An inspection of Table I will show that in the case of the parlor 26 of the contestants have used registers whose width is the same as the diameter of the connecting pipe, while 7 have used registers that are narrower and 7 that are wider than the diameter of pipe.

The tables are not as complete as they would have been if the authors of a number of the essays had not omitted putting the size of registers and pipes on the plans submitted.

As an example of size of pipe required to convey the hot air to any of the upper rooms, the northeast chamber is taken, although in Table II the sizes of pipes and registers are given, as far as obtainable, for the four bedrooms. An inspection of the table will show that pipes are used varying in size from 3½ x 8, 3 x 17, 4 x 16 to 5½ x 14. In connection with these pipes registers are used varying in size from

---

* Reprinted from *The Metal Worker*, November 16, 1889.

8 x 10 to 10 x 14. In one instance two 6 x 8 registers have been used in connection with 4 x 6 inch flues, this being the only instance where two pipes and registers are used to heat one of the bedrooms. While it is presumed that the tables will be interesting to many, it is also hoped they will be of service to those who have had little experience in the furnace business or have been so situated as to have had few opportunities of seeing how work is done by experts.

In case a tinner of limited experience in furnace work was called upon to determine the sizes of pipes and registers necessary to be used in connection with the heating of a house similar to the one of which plans were published in *The Metal Worker*, by looking over the tables he could see at a glance the sizes of pipes or registers the various authors have considered to be of the proper size. It will be noticed that quite a number of the contestants have indicated that 3½ x 13 or 14-inch pipe in connection with 9 x 12 or 10 x 14 registers were the proper sizes to use. While the thickness of most of the pipes is given at 3, 3½ or 4 inches, it might be presumed that the pipes are intended to be made such a thickness that they can be placed in partitions made of 2 x 4-inch studding, the space between the lath sometimes being less than 4 inches, depending on how much is taken off in truing up. There is no particular method of designating the sizes of partition pipes, as one person may give the size of the inner pipe and another the size of the outer. Also in some instances the authors of the essays have used single pipes in the partitions, covering the studding with tin and using iron lath, while others have used double pipe and in some cases the pipes are to be covered with asbestos paper. The intention in presenting the tables is not to anticipate the essays that are to be published, but to give a bird's-eye view of the sizes shown, which could only be done by the use of a table or similar method. An interesting use of the tables may be mentioned. Let the reader follow down any particular column and notice the variation in sizes, and then follow through on any line from left to right, noting the variation, if any, there may be in the sizes of registers or pipes used for the various rooms, and then follow down any particular column and note the size of register or pipe that is intended for any particular room.

Number of essay.	Parlor. Size of Pipe.	Reg.	Library. Size of Pipe.	Reg.	Dining-room. Size of Pipe.	Reg.	Hall. Size of Pipe.	Reg.
	Inch.	Inch.	Inch.	Inch.	Inch.	Inch.	Inch.	Inch.
1	14	14 x 18	12	12 x 16	12	12 x 16	12	12 x 16
2	o	12 x 15	o	12 x 15	o	12 x 15	o	12 x 15
3	12	12 x 15	10	10 x 14	10	10 x 14	12	12 x 15
4	10	12 x 15	10	12 x 15	10	12 x 15	12	12 x 15 2
5	10	12 x 14	10	10 x 14	10	10 x 14	10	10 x 12
6	10	10 x 14	10	10 x 14	10	10 x 14	10	10 x 14
7						Not given.		
8	o	12 x 14	o	12 x 14	o	12 x 14	o	12 x 15
9								
10	12	10 x 14	10½	10 x 14	10½	10 x 14	o	10 x 14
11	o	12 x 17	o	12 x 15	o	12 x 15	o	12 x 17
12	12	12 x 15	12	12 x 15	12	12 x 15	12	12 x 15
13	11	10 x 14	12	12 x 15	11	10 x 14	10	10 x 12
14	12	o	12	o	12	o	10	o
15	10	10 x 14	10	10 x 14	10	10 x 14	12	12 x 15
16	12	12 x 15	12	12 x 15	12	12 x 15	12	12 x 15
17	11	12 x 15	10	10 x 14	10	10 x 14	11	12 x 15
18	o	10 x 14	o	10 x 14	o	10 x 14	o	12 x 15
19	12	12 x 15	12	12 x 15	12	12 x 15	12	14 x 22
20	o	10 x 14	o	10 x 14	o	10 x 14	o	12 x 15
21	10	10 x 16	10	12 x 16	10	12 x 16	12	12 x 19
22	11	10 x 14	11	10 x 14	11	10 x 14	9	10 x 14
23	o	12 x 15	o	12 x 15	o	12 x 15	o	12 x 15
24	10	10 x 14	10	10 x 14	10	10 x 14	10	10 x 14
25	12	10 x 14	10	10 x 14	10	10 x 14	12	12 x 16
26	10	10 x 14	10	10 x 14	10	10 x 14	12	12 x 15
27	2 4½ x 9	2 6 x 10	2 4½ x 10	2 7 x 10	2 4½ x 9	2 6 x 10	4 4½ x 8	o
28	10	10 x 14	10	10 x 14	10	10 x 14	10	10 x 14
29	12	12 x 17	12	12 x 17	10	12 x 15	14	14 x 18
30	12	12 x 15	12	12 x 15	12	12 x 15	12	12 x 15
31	9	10 x 14	9	10 x 14	9	10 x 14	11	12 x 16
32	9	10 x 14	9	10 x 14	9	10 x 14	10	12 x 15
33				Sizes not	shown	on plans		
34				Sizes not	shown	on plans		
35	o	12 x 15	o	12 x 15	o	10 x 12	o	16 x 22
36	o	10 x 14	o	10 x 14	o	10 x 14	o	10 x 14
37	10	10 x 14	10	10 x 14	10	10 x 14	10	10 x 14
38	10	10 x 14	10	10 x 14	10	10 x 14	12	12 x 15
39	10	9 x 14	10	9 x 14	10	9 x 10	10	9 x 14
40	10	12 x 16	10	12 x 15	10	12 x 15	10	12 x 15 2
41	11	10 x 16	11	10 x 16	11	10 x 16	13	10 x 12
42	12	12 x 14	12	12 x 14	12	12 x 14	12	12 x 14
43	10	12 x 15	10	12 x 15	10	12 x 15	10	12 x 15
44				Sizes not	shown	on plans Pedestal		
45	o	12 x 15	o	10 x 16	o	16 x 21	o	14 x 14
46				Sizes not	shown	on plans		
47	12	10 x 14	12	10 x 14	12	10 x 14	12	12 x 14
48	10	10 x 14	10	10 x 14	10	10 x 14	10	10 x 14
49	12	12 x 15	12	12 x 15	12	12 x 15	10	10 x 14
50				Sizes not	shown	on plans		
51	10	10 x 14	10	10 x 14	10	10 x 14	12	14 x 22
52	10	10 x 16	9	9 x 12	9	9 x 12	10	10 x 16
53				Sizes not	shown	on plans		
54	10	10 x 14	10	10 x 14	10	10 x 14	12	10 x 14
55	12	12 x 15	12	12 x 15	12	12 x 15	12	12 x 15
56				Sizes not	shown	on plans		
57				Sizes not	shown	on plans		
58	12	12 x 15	12	12 x 15	12	12 x 15	10	2 10 x 12
59				Sizes not	shown	on plans		
60				Sizes not	shown	on plans		
61	12	12 x 15	12	12 x 15	12	12 x 15	14	12 x 19
62	10	10 x 14	10	10 x 14	10	10 x 14	10	10 x 14

Note.—o indicates that no size is shown on plans.

Table I.—Sizes of Heating Pipes and Registers.—First Floor.

Number of essay.	N. E. Chamber.		S. E. Chamber.		S. W. Chamber.		N. W. Chamber.	
	Size of Pipe.	Reg.	Size of Pipe.	Reg.	Size of Pipe.	Reg.	Size of Pipe.	Reg.
	Inch.	Inch.	Inch.	Inch.	Inch.	Inch.	Inch.	Inch.
1	3½ x 12	10 x 14	3½ x 12	9 x 12	3½ x 9	9 x 12	3½ x 12	9 x 12
2	o	o	o	o	o	o	o	o
3	4 x 13	9 x 14	4 x 11	8 x 12	4 x 11	8 x 12	4 x 13	9 x 14
4	o	10 x 14	o	10 x 14	o	10 x 14	o	10 x 14
5	o	10 x 14	o	10 x 14	o	10 x 14	o	10 x 12
6	3½ x 12	9 x 12	3½ x 12	9 x 12	3½ x 12	9 x 12	3½ x 10	8 x 12
7								
8								
9								
10	o	8 x 12	o	8 x 12	o	8 x 12	o	8 x 12
11	o	10 x 14	o	9 x 14	o	9 x 14	o	8 x 12
12	3½ x 12	10 x 12	3½ x 12	10 x 12	3½ x 12	10 x 12	3½ x 12	9 x 12
13	o	8 x 10	o	8 x 12	o	8 x 18	o	6 x 10
14								
15	3 x 12	9 x 12	3 x 12	9 x 12	3 x 12	9 x 12	3 x 12	9 x 12
16	3½ x 12	10 x 12	3½ x 12	10 x 12	3½ x 12	10 x 12	3½ x 10	8 x 12
17	3½ x 12	8 x 12	3½ x 12	8 x 12	3½ x 12	8 x 12	3½ x 12	8 x 12
18	o	8 x 12	o	8 x 12	o	8 x 12	o	8 x 12
19	o	10 x 14	o	10 x 14	o	10 x 14	o	10 x 14
20	o	10 x 14	o	10 x 14	o	10 x 14	o	
21	2½ x 12	9 x 12	3½ x 12	9 x 12	3 x 12	9 x 12	3 x 10½	8 x 10
22	3½ x 8	8 x 10	3½ x 8	8 x 10	3½ x 8	8 x 10	o	8 x 10
23	o	8 x 10	o	8 x 10	o	8 x 10	o	8 x 10
24	4 x 14	10 x 14	4 x 14	10 x 14	4 x 14	10 x 14	4 x 14	10 x 14
25	3½ x 15½	10 x 12	3½ x 16½	8 x 12	3½ x 15½	10 x 12	3½ x 10	8 x 12
26	o	9 x 12	o	9 x 12	o	9 x 12	o	8 x 12
27	2	2	2	2	2	2	2	2
	4½ x 6	6 & 8	4½ x 6	7 & 8	4½ x 6	6 & 8	4½ x 5	6 & 8
28	3½ x 12	9 x 12	3½ x 12	9 x 12	3½ x 12	9 x 12	3½ x 12	9 x 12
29	5 x 13	10 x 14	5 x 13	10 x 14	5 x 13	9 x 12	4 x 14	9 x 12
30	o	10 x 10	o	10 x 10	o	10 x 10	o	8 x 10
31	5 x 10	10 x 14	5 x 10	10 x 14	5 x 10	10 x 14	5 x 10	8 x 12
32	3½ x 13	8 x 12	3½ x 13	8 x 12	3½ x 13	8 x 12	3½ x 13	8 x 12
33			Sizes not	shown	on plans.			
34			Sizes not	shown	on plans.			
35	o	12 x 15	o	12 x 15		10 x 12	o	o
36	o	9 x 12	o	9 x 12		9 x 12	o	9 x 12
37	o	8 x 12	o	8 x 12		8 x 12	o	8 x 12
38	3 x 10½	9 x 12	3 x 10½	9 x 12	3 x 10½	9 x 12	3 x 10½	8 x o
39	o	8 x 12	o	8 x 12	o	8 x 12	o	8 x 12
40	4 x 16	10 x 14	4 x 16	10 x 14	4 x 16	10 x 14	4 x 16	10 x 12
41	o	10 x 12	o	10 x 10	o	10 x 14	o	10 x 10
42	3 x 17	10 x 14	3 x 17	10 x 14	3 x 17	10 x 14	3 x 17	10 x 14
43	3 x 12	9 x 12	3 x 12	9 x 12	3 x 12	9 x 12	o	
44			Sizes not	shown	on plans.			
45	o	10 x 14	o	10 x 14	o	10 x 14	o	10 x 12
46			Sizes not	shown	on plans.			
47	4 x 12	8 x 10	4 x 12	8 x 10	4 x 12	9 x 12	4 x 12	8 x 10
48	3⅝ x 12⅝	9 x 12	3⅝ x 12⅝	9 x 12	3⅝ x 12⅝	9 x 12	3⅝ x 12⅝	8 x 10
49	o	9 x 12	o	9 x 12	o	9 x 12	o	10 x 14
50			Sizes not	shown	on plans.			
51	o	10 x 12	o	10 x 12	o	10 x 12	o	10 x 12
52	o	8 x 10	o	8 x 10	o	8 x 10	o	8 x 10
53			Sizes not	shown	on plans.			
54	o	9 x 14		10 x 12		9 x 14		8 x 12
55		10 x 14		10 x 14		10 x 14		10 x 14
56			Sizes not	shown	on plans.			
57			Sizes not	shown	on plans.			
58	5½ x 14	10 x 12	5½ x 14	10 x 12	4 x 14	10 x 12	4 x 14	9 x 12
59			Sizes not	shown	on plans.			
60			Sizes not	shown	on plans			
61	3½ x 13	10 x 12	3½ x 13	10 x 12	3½ x 13	10 x 12	3½ x 13	9 x 12
62	4 x 12	8 x 12	4 x 12	8 x 12	4 x 12	8 x 12	4 x 14	8 x 12

Note.—o indicates that no size is shown on plans.

Table II.—Sizes of Heating Pipes and Registers.—Second Floor.

COMPARISONS OF ESTIMATES.*

Not the least interesting point to be observed in looking over the various essays submitted in *The Metal Worker* hot air heating contest is the difference existing in the estimates of cost. It is hardly to be expected that the various estimates would be nearly alike, as the contestants were allowed so much latitude in regard to the manner in which the work was to be done. As an ideal system of heating and ventilating was to be shown for the house of which plans were published, it may be that some of the contestants have done the work in a different manner from what they would if figuring on a job that was to be let in the usual manner, where the lowest bid would be supposed to receive the most attention. In this connection the following extract from the advertisement published in *The Metal Worker* of December 15, 1888, is worthy of consideration : •

"There is no limit as to the expense of the systems of heating to be shown. The contestants are to assume that the dwelling is to be built and furnished by a gentleman of ample means, who stands ready and willing to pay up to any amount for which there is value received. On the other hand, he is a careful business man and has no money to throw away. He is not willing to pay $500 for work that should be done for $400, nor does he propose in this case to pay for anything that is merely ornamental in character, having no useful function. The house in all its appointments is to be plain and substantial, and the heating system is to correspond. This said, however, he wants the best in every respect that the present state of the art affords. He is willing to pay for an ideal system of heating of the best material and workmanship. This also makes it necessary that reasons for the use of different features of apparatus should be fully set forth."

As the supposed builder is a careful business man, it is only natural that he should look over the various estimates of cost and perhaps prepare a table as we have done. In Table III the estimate

* Reprinted from *The Metal Worker*, October 12, 1889.

No. of essay.	Estimate of cost.	Cost of furnace.	Tin work. Cost of pipes, &c.	Cost of ventilat'rs and rezisters.	Kind of regulator.	Remarks.
1						First prize.
2						No estimate of cost.
3	$370.23	$150.00	$115 78	$40.75		
4	371.15	125.00	52.35	38.19		Regulator, $40.
5	409.70	140.00	97.25	78.75		
6	302.30	205.00	126.10	25.00		
7	208.00	125.00	75 00	50.00		
8	450.00	155.00	180.00	51.64		
9						No estimate of cost.
10	521.83	125.00	100.00	46.23		
11						No estimate of cost.
12	358.25	125.00	130.00	41.32		
13	479.00	250.00	92.50	53.00		
14	518.48	175.00	113.80	47.10		
15	346.93	130.00	112 75	38.08		
16	336.50	135.10	133.60	27.00		
17	420.00	175.00	110.45	42.95		
18	302.30	175.00	72.37	24.23		
19	325.00	168.60	60.50	46.50		
20	458.00	182.50	70.20	134.75		Ventilators, $50.
21	325.00	218.00	113.10	27.00		
22	257.00	125.00	75.35	23.25		
23	306.67	150 00	101 00	62.23	Chain and damper	
24	401.15	213.00	78.00	26.25		
25	346.15	185.00	91.14	95.75		
26	288.00	130.00	75.39	42.77	Chain and damper	
27	783.00	310.00	148.00	86.00		
28	259.00	111 50	98.00	15.50		
29	544.95	195.00	161.00	36.50		
30	449.86	75.00	152.65	59.00	Chain and damper	
31	804.32	435.60	178.50	55.22	Electric, $40.	
32	635.00	300.00	137.10	47.40	Electric, $40.	
33	422.00	209.40	132.39	31.15	Elec. if wanted.	
34						No estimate of cost.
35	375.70	200.00	71.30	69.00	Chain and damper	
36	418.66	125.00	129.08	54.14		
37	273.70	115.00	75 90	26.55		
38	269.50	145.00	73.10	32.65		
39	507.22	173.53	96.88	76.40		
40						Second prize.
41	704.77	186.70	117.62	41.60	Chain and damper	
42	492.75	187.50	150.00	33.25		
43						No estimate of cost.
44	713.04	285.00	185 33	79.45		
45	402.50	154 50	90 65	45 56		
46	502.50	175.00	150.00	50.00		
47	559.27	175.00	116.77	23.00	Reguit'r, price $6	
48	399.25	175.00	134 05	29.80		
49	449.93	208.50	90.00	67.20	Regulator, $25.	
50	592.09	150.00	107.87	37.67		
51	375.94	175.00	73.43	39.94		
52	604.90	273.25	175.00	73.27		
53	355.99	122.50	100.30	48.60		
54	285.00	125.00	84.05	52.60		
55	753.74	150.06	110.70	103.33		
56	482.87	295.00	124.80	34.07		
57	547.50	175.00	237.50	56.00		
58					Pat. reg., $15.	No estimate of cost.
59	621.00	109.00	160.81	44.95		
60	365.00	190.00				Price of vent. and reg. not given separate.
61	447.90	149.00	159.20	19.40		
62	400.64	140.00	116.77	30.89		
Average cost	$443.13	$186.70	$112.87	$46.46		

Table III.—Estimates in "The Metal Worker" Furnace Competition.

for the job is given, then follow the cost of furnace, cost of tin work, then the cost of the registers and ventilators.

As this table is a bird's-eye view of the various estimates, it would be difficult to show the details that go to make up the amounts, since there are a number of items in each estimate that are not included in the tables. It is certainly interesting for one to see how the contestants have looked at the subject from various points of view, and what a wide difference there is in the results. Regarding the subject of estimates, it might be well to make another extract from the original advertisement, which is as follows :

"With each essay in each of these contests a detailed estimate of materials and workmanship is to be presented, by which we mean a complete list of items of material and the labor to put same in place, showing in the aggregate the cost of the system to the house owner. A lump estimate will not be deemed sufficient."

From the above it would appear that the estimate is to cover all work necessary " to put the same in place."

We would mention right here that the numbers by which the essays are referred to are arbitrary markings of our own and are not in any way based upon merit.

By looking at the first column, which shows the estimates of cost of apparatus, it will be seen that No. 31, which is the highest, considers $806.32 as the figure at which the job is to be done. In this estimate are a number of items that some would consider as extras, or work that should be done by others than the heating contractor, as—

Asbestos wrapping	$18.00
Carpenter work	14.00
Excavation	5.00
Flagstone for cold air duct	11.00
Electric heat regulator	40.00
Air filter	40.00
Total	$128.00

The above items are taken from the estimate to show some of the parts that go to make up the whole amount that are not included in the table. The price of furnace, which is $435, indicates that it is to be of superior quality and size ; the cost of brick casing, which is given at $60, is included in the above amount.

The estimate for tin work in this case, also for the registers and ventilators, is below some others, as can be seen by reference to the table.

The next highest estimate is No. 27, being $783.   As there is no
particular mention made, excepting as below, of the work of putting
in the apparatus, it is presumed it is included in the estimate. The
price of furnace, $310, includes the setting in brick, &c.   Among the
items that some would consider as extras are :

Extra masonry in chimneys	$20.00
Unforeseen expenses, &c	47.00
Salary of superintendent	84.00
Total	$151.00

While the last item may be unusual, judging from the fact that it
does not appear in any other estimate, it is probable that furnace men
would like to see it on every estimate.

The lowest estimate is given by No. 22, being $257 for the whole.
In this is an allowance of $15 for labor and $13 for cold air box and
pit. An inspection of the second column in the table will show that
there are a number of estimates in which the price of the furnace is
greater than in the above, as are the prices for tin work and ventilators and registers. That is, there are estimates in which each of the
items given in the table is lower than in this estimate, though it is
the lowest of the estimates.

The difference in price between No. 31, the highest, and No. 22,
the lowest, is $549.32, which certainly shows a diversity of opinion re-
garding what is necessary to heat a house properly.

The next lowest estimate is by No. 28 ; in this the total cost is
$259. The furnace is placed at $111.50, which is lower in price than
in No. 22, while the price of tin work is higher. The price of registers, as carried out in the estimate, is as follows :

4 10 x 14 registers	$4.00
7 9 x 12 registers	6.00
2 8 x 10 registers	1 00
4 10 x 14 borders	2.00
4 4 x 10 ventilators	2.50
Total	$15.50

That a mistake has been made in the above is quite evident, for
one could hardly furnish 17 registers and 4 frames for $15. In case
the estimate had been made for a bid and the job taken at the above
figures, it is probable the profits would not have been as great as an-
ticipated.

The next lowest estimate is No. 38, the price being $269.50. In this estimate there are no striking features other than are shown in the table.

AVERAGE COST.

The average of the 54 estimates of cost is shown at the foot of the table, that for the entire work being $443.13. The estimates nearest this amount are No. 61, $447.90; No. 30, 449.86, and No. 49, $449.93. In No. 61 the registers are placed at $19.40, which amount is below many others. In No. 30 there is one rather startling feature as compared with others, and that is the price of the furnace, which is placed at $75, while the registers and ventilators are $50, being two-thirds of the price of the furnace.

AVERAGE AND PRICE OF FURNACES.

In the second column is shown the cost of furnaces as derived from the various estimates, the cost including the casing or brick covering, as the case may be. The highest priced furnace is found in No. 31, it being $435, while in No. 30 is the lowest-priced, which is $75. The average price of the furnaces is $186.70, as shown at the bottom of the table, and the furnace nearest in price to the average is No. 42, which is placed at $187.50.

COST AND AVERAGE OF TIN WORK.

In determining the cost of tin work from the estimates it has been difficult to obtain exact figures, as in some instances the item of work has been stated in such a manner as to make it difficult to determine if the allowance was for making the pipes, &c., putting them in place, or for both, yet this column of the table, while rather imperfect, is quite interesting as showing to a certain extent the differences of opinion on the subject. It would be about impossible to make an explanation of the cost of tin work without publishing that part of the estimates and the accompanying plans, as the location of the furnace and registers, method of heating upper rooms and other causes have much to do with this part of the subject.

The highest estimate for tin work is found in No. 57, being $237.50. Of this amount $103.75 is for ventilating pipes, &c., leaving $133.75 for the furnace work. The next highest estimate is in No. 44, the work being placed at $185.33, of which $82.98 is for ventilating pipes, &c., leaving $102.35 for furnace work. The lowest estimate for this branch is found in No. 4, the price being $52.35. In this estimate there is no mention made of ventilating pipes. The next lowest estimate is in No. 19, the price being $60.50.

288   THE METAL WORKER ESSAYS.

The average estimate of tin work is $112.87, and the one coming nearest to this is No. 15, which is $112.75, and the one next in order is No. 21, at $113, which is 13 cents above the average. In only one are ventilating pipes mentioned, and then at a cost $13.

REGISTERS AND VENTILATORS.

Regarding the cost of registers and ventilators, much would depend upon the finish, size and system of ventilation employed, &c. As there are a certain number of rooms in the house to be heated, and

Estimates of total cost.	No.		Tin work, &c.	No.	
Highest estimate	31	$806.32	Highest estimate	57	$237.50
Two next highest	{ 27	783.00	Next highest	44	185.33
	{ 55	763.74	Lowest estimate	4	52.35
Lowest estimate	22	257.00	Next lowest	20	70.20
Two next lowest	{ 38	269.50	Average price		112.87
	{ 28	259.00	Nearest to average	15	112.75
Average cost		444.13	Near to average	21	113.00
	{ 49	449.93			
Nearest to average	{ 30	449.10			
	{ 61	447.90			

Furnace only.	No.		Registers and ventilators.	No.	
Highest price	31	$435.00	Highest price.	20	$134.75
Next to highest	27	310.00	Next to highest	55	103.33
Lowest price	30	75.00	Lowest price	28	15.50
Next to lowest	28	111.00	Next lowest	61	19.40
Average price		168.70	Average price		46.48
Nearest to average	42	187.00	Nearest to average	19	45.50

Table IV.—Comparisons of Estimates.

it is supposed a register will be used in each room, there is at least one point of certainty connected with the price.

The highest estimate of cost for the registers and ventilators is found in No. 20, the price being $134.75, and composed of the following items :

17 convex registers and borders, at $4 .... $68.00
1 12 x 15 floor register and border, nickel plated......................... 7.00
1 10 x 14 floor register and border, black................................ 3.75
8 11 x 16 circular top ventilators, bronze finish......................... 56.00

Total.. $134 75

www.ingramcontent.com/pod-product-compliance
Lightning Source LLC
Chambersburg PA
CBHW021512210326

41599CB00012B/1231